GREEN OR BUST

GREEN OR BUST

James Wilkinson

BBC BOOKS

James Wilkinson has worked for the BBC since 1975, first as Radio's Science and Aviation Correspondent and, since 1982, as TV's Science and Medical Correspondent. He has written two previous books, *The Conquest of Cancer* and *Tobacco: The Truth Behind the Smoke-screen.*

The author and publishers wish to thank Denise Byers for permission to quote copyright material from *State of the World 1990*, Worldwatch Institute, Washington DC, USA.

Published by BBC Books,
a division of BBC Enterprises Limited,
Woodlands, 80 Wood Lane, London W12 0TT
First published 1990

ISBN 0 563 36032 1

Set in 10/12 pt Palatino by Ace Filmsetting Ltd, Frome
Printed and bound in England by Clays Ltd, St Ives plc
Cover printed by Clays Ltd, St Ives plc

Contents

Acknowledgements

I would like to acknowledge the help given to me by the following people who read relevant parts of the manuscript and made many helpful suggestions. The emphasis, though, and any mistakes, are mine alone: Dr Adam Brown of the Energy Technology Support Unit at Harwell, Dr John Gittus of the British Nuclear Forum, Dr Neil Cape of the Institute of Terrestrial Ecology's Edinburgh Research Station, Dr Geoff Williams of the British Geological Survey, Dr Mick Kelly of the University of East Anglia Climate Research Unit, Dr Daniel Osborn of the Institute of Terrestrial Ecology, Monks Wood Experimental Station, Celia Kirby of the Institute of Hydrology, Dr Tony Cox of the Natural Environment Research Council, Dr Francis Sullivan of the World Wide Fund for Nature and Dr Philip Williamson of the Plymouth Maritime Laboratory. I am also grateful to Mr Jeremy Baldwin of the Natural Environment Research Council for his help.

Publisher's Note:
The author has used both metric and imperial measurements throughout the text in order to quote accurately the source of his information.

CHAPTER ONE

The Ailing Earth

The environment touches everyone. The air we breathe, the food we eat, the water we drink and bathe in, the countryside we walk in – all these are affected in one way or another by mankind's polluting activities. As the results of these activities come to light and the pressures on the environment increase, so does the concern with which we view the world about us. With the recent boom in media exposure to environmental problems it would be easy to think that things have never been worse. Yet in some ways our domestic environment in Britain is much better than it has been in the past: the poisonous fumes of Victorian industrial Britain are now, thankfully, present only in faded photographs and etchings. Elsewhere in the world, however – notably in some places in Eastern Europe – the story is quite different. In East Germany and Poland, for example, air pollution can be tasted, and water and land pollution are killing people. But even in these places, the new awakening is leading to economic changes which it is hoped will bring about a cleaner environment.

But although industrial pollution has been recognised, if not entirely dealt with, it has been replaced by new threats. The increased use of the motor car has brought in its wake a host of air pollution problems, together with the threat of cities locked in everlasting traffic jams.

And the biggest threats of all, the greenhouse effect and the hole in the ozone layer, which could between them cause cataclysmic changes in the world, are probably more important than any previously recognised environmental hazard.

Concern over the environment became a focus of political

debate two decades ago with the formation of specific pressure groups like Friends of the Earth and Greenpeace, even though, particularly in Britain, certain measures, notably the Clean Air Act, had anticipated more general environmental concerns by several decades.

Friends of the Earth began in the United States in the late sixties as an offshoot of the Sierra Club, one of the largest US conservation groups. In 1971 it was officially launched in Britain. Its first stunt took place at the London headquarters of Schweppes where 2000 non-returnable bottles were deposited. The protest received considerable media coverage and the organisation grew rapidly. One of its first major political battles concerned the plans by the mining giant Rio Tinto Zinc to open a huge open-cast copper mine in the heart of the Snowdonia National Park. After two years of campaigns the company withdrew its plans.

Greenpeace began as an international organisation in 1971 when a group of American and Canadian environmentalists chartered a boat to protest at proposed US nuclear testing in the Aleutian Islands off Alaska. There followed campaigns to save dwindling species such as whales and seals which brought the plight of these creatures to international attention. Greenpeace opened its London office in 1977 and the campaigning has continued ever since.

In 1972 the United Nations set up a seminal **Conference on the Human Environment** in Stockholm. The meeting was attended by 1200 delegates from more than 100 nations, and it was here that the need for concerted international action to protect the environment found its voice. In summing up, the Canadian Secretary General of the conference, Mr Maurice Strong, listed three main areas of concern: water supplies, ocean pollution and the urban crisis. In a rallying call he told the delegates that they were embarking on a 'journey of hope'. 'In the decades ahead we must learn to conquer our own divisions, our greeds, our inhibitions and our fears or they will conquer us.'

There had been concern that the conference would split between the rich and the poor. The developing countries feared they would be told that they would have to make sacrifices in order to preserve an environment already poisoned by the industrialised world. And there was the added irony that the conference was being held at the time of the Indo-China war in which America was destroying huge tracts of land with bombs and defoliants.

In the event, the conference concluded with a ringing declaration which heralded at last an international recognition that spaceship earth needs nurturing and cherishing if it is to survive – and us with it. A 'Declaration on the Human Environment' was drawn up, outlining standards and principles to guide nations in matters relating to the environment.

The conference also approved 100 recommendations which were detailed in an 'Action Plan', divided into three parts:

The first was the establishment of a global assessment programme called Earthwatch which was to identify and measure environmental problems and warn of any impending crisis.

Secondly, there were management activities to put in place to implement measures to protect the environment.

Thirdly, the action plan would support educational, financial and organisational matters necessary to safeguard the environment.

During the conference a number of speakers raised the question of the effects of weapons on the environment, the testing of nuclear weapons and the use of chemical weapons. The Swedish Prime Minister Mr Olof Palme attacked the indiscriminate bombing and use of defoliants as an outrage bordering on ecocide. He made a plea for ecological warfare to cease immediately. His remarks brought a sharp reaction from the United States to whom he was referring.

A resolution to stop the atmospheric testing of nuclear weapons was passed in Plenary session but the French, who were planning a series of atmospheric tests in the Pacific, said they would not be bound by it.

One of the first recommendations to be adopted called for control over the release of toxic metals and chemicals into the environment. The Japanese gave an example of how pollution by modern chemicals can cause illness. Earlier that year 1081 people had become ill and 16 had died after being poisoned by polychlorinated biphenyl which had leaked into rice bran oil which many Japanese use for cooking. (PCBs had already by that time been banned in Canada.) A resolution urged governments to use the best practicable means to minimise the release of toxic or dangerous substances, especially persistent chemicals like mercury and other metals and organochlorine pesticides.

The ground work was prepared for a conference on ocean dumping to be held in Britain later (see Chapter 9). The conference also called for a ten-year ban on commercial whaling, though Japan, which with the Soviet Union accounted for 80 per cent of the world catch, objected.

The subject of over-population was raised and UN bodies like the World Health Organisation were urged to help governments with their family planning needs.

The conference for the first time made concern for the environment truly international. Looking back on it eighteen years later it is fascinating to see how some of the issues raised then have developed. There is certainly far more international awareness now of the need to protect the environment. Issues like the protection of whales and the control over ocean dumping have been tackled strenuously. And many of the initiatives which began then have grown and borne fruit. But equally it can hardly be said that the world is now a cleaner and healthier place as a result of what was discussed there. The momentum which has given rise to worldwide pollution will take a very long time indeed to slow down.

Perhaps most fascinating of all is the fact that two of the most crucial environmental issues now facing humankind – the greenhouse effect and the destruction of the ozone layer by CFCs – were not even mentioned at the conference. They have crept up on us almost unnoticed and will affect every one of us.

Nowadays there are plenty of organisations which are quick to spot a potential new environmental threat and to stimulate political action. But back in 1972 environmental organisations like Friends of the Earth were still in their infancy and regarded by many as idealists with their feet off the ground. How things have changed in the intervening years. The environment now comes at the top of the political agenda in countries throughout the world.

Among the significant follow-up meetings and commissions established since Stockholm was **The Brundtland Commission** – the World Commission on the Environment, established in 1984 by the General Assembly of the United Nations and chaired by the Norwegian Prime Minister, Mrs Gro Harlem Brundtland. The Commission was set up to propose long-term environmental strategies for 'sustainable development', a phrase which has become integrated into the environmental jargon. It was also to recommend ways in which concern for the environment might be translated into greater cooperation between industrial and developing countries and to look at long-term environmental issues.

In 1987 the Brundtland Commission published its final report entitled 'Our Common Future'. The report was another landmark in the international struggle for a cleaner, safer world. It concentrated on issues like the security of the food supply, the loss of species and the importance of genetic diversity, energy, industry and human settlements. It strongly recommended that governments review their programmes in areas such as forestry and agriculture and settlements where there might be a threat to natural habitats. It urged the promotion of energy efficiency and more investment in renewable sources of energy. Developing countries were encouraged to strengthen their controls on pollution-intensive resource-based industries. And it argued strongly that investment agencies should incorporate sustainable development criteria into their policies.

On what it called the 'urban challenge' it pointed out that in just 5500 days (15 years) the developing countries would have to increase their capacity to produce and manage their urban infrastructure, services and shelter by 65 per cent in order to maintain even existing conditions. And it drew attention to the dilemma over spending on arms: 'Governments and international agencies should assess the cost effectiveness, in terms of achieving security, of money spent on armaments compared with money spent on reducing poverty or restoring a ravaged environment.'

Internationally, concern for the environment has gained a strong political foothold with the Green Party in certain European countries, notably in Germany and Sweden, where it is now quite a force. In Britain, despite the activities of organisations like Greenpeace and Friends of the Earth, environmental consciousness was latent until it was forced to the top of the political agenda by none other than Mrs Thatcher, who, until her speech to the Royal Society in September 1988, had been a particular target for environmentalists.

It was the first time she had personally shown her concern for the environment. She pointed out that medical advances had saved millions of lives, causing a dramatic rise in population; agricultural research had led to increasing use of fertilisers which caused pollution from nitrates, and an increase in the use of fossil fuels had produced carbon dioxide. 'For generations', she said, 'we have assumed that the efforts of mankind would leave the fundamental equilibrium of the world's systems and atmosphere stable. But it is possible that with all these enormous changes . . . concentrated into such a short period of time, we have unwittingly begun a massive experiment with the system of this planet itself.'

She highlighted the potential problem of global warming, the existence of a hole in the ozone layer and acid rain. At a stroke she transformed the climate of opinion on the environment in this country. From being a subject which the opposition parties could use to attack her, that particular platform was virtually cut from beneath their feet. Now all they could do was claim they were even greener than she was. Environmental organisations found themselves acknowledging her role in raising the environmental consciousness of the nation. In some ways she had been too successful because, having raised the temperature, she perhaps sparked off a greater expectation in the electorate than she could satisfy and by doing so gave an opportunity for Britain's own Green Party to become fully established. **The Green Party** achieved unprecedented success in the European elections in 1989, attracting 2 million votes – 15 per cent of the vote. In some ways, though, its success was to be short-lived. It relied too heavily on support which was perhaps not as soundly based as it had hoped. Many people voted for the Green Party to send a signal to the Government that they really were concerned about the environment. Aspects of the Green Party's manifesto which perhaps had not been fully appreciated by voters in the euphoria of the green crusade were the commitment to unilateral disarmament, withdrawal from NATO and the European Community, what has been described by some as their 'no growth' policy, and the quirky suggestion that the Green Party should have no single leader. Since 1982 the party has had a goal to reduce Britain's population by 15–20 million by such things as education and a changed attitude to childlessness. It is unlikely to repeat its success in the next election. For a while, though, its political impact was considerable. Its party conference in September 1989 attracted almost more journalists than delegates. It had to limit the number of press passes to 500. The highest number of journalists to attend its previous conferences was 12.

Mrs Thatcher's apparent commitment to the environment gave rise in December 1989 to the Government's **Environmental Bill** – 150 pages of legislation which would tighten controls over a whole range of diverse issues, from new ways to measure pollution from factories to controls on the release of genetically engineered organisms. The Bill dealt with a whole lot of measures which had accumulated over the previous decade, it introduced the concept of integrated pollution control which means that the effects of polluting emissions from factories on air, water and land will have to be considered together rather than separately as had previously been the case. The Bill even dealt with the mundane subject of litter. Perhaps its most controversial clauses concerned the splitting up of the Nature Conservancy Council into separate agencies for Scotland, Wales and England.

Recently, the environment has become an extremely emotive issue, and not just in Britain. Major catastrophes like the Chernobyl disaster in the Soviet Union have served to stress just how small the planet is and how delicately balanced the ecosystem is.

Concern for the environment is reflected in the increasing quantity of legislation in the European Community in the eighties. In 1976 the Community agreed eight environmental measures. In 1980 there were 13, in 1984 23 and in 1988 there were 30.

Yet if development is to be sustainable, then implementing measures to protect bits of the environment in a piecemeal fashion is only going half way to solving the problem. What is needed is a more fundamental review of the forces which underlie the destruction of the environment and a more practical method of tackling them at source.

One such review was produced in 1989 by Professor David Pearce, Professor of Economics at University College, London, under a research contract from the Department of the Environment. It put forward the idea that to conserve the environment it must have a monetary value which has to be integrated into the costings of any endeavour which has an impact on the environment. To quote from the preface to his report: 'Sustainable development is feasible. It requires a shift in the balance of the way economic progress is pursued. Environmental concerns must be properly integrated into economic policy from the highest to the most detailed level. The environment must be seen as a valuable, frequently essential input to human wellbeing. Sustainable development means a change in consumption patterns towards more environmentally benign products, and a change in investment patterns towards augmenting environmental capital.' In short everything has its value and that value should be reflected in the price we have to pay for goods and services. As Professor Pearce says: 'If one generation leaves the next generation with less wealth, then it has made the future worse off. But sustainable development is about making people better off – hence [it is] a policy which leaves more wealth for future development.' Mrs Thatcher summed it up by saying that we needed to maintain a 'fully repairing lease' on the world. At present no country in the world is doing that.

CHAPTER TWO

The Greenhouse Effect

Of all the many worries about the effect mankind is having on the environment there is one above all others which gives most cause for concern: the greenhouse effect. Surprisingly, although it is now recognised as the greatest threat to the future of our planet, in 1972, at the Human Environment conference in Stockholm, climate change was not even listed as one of the threats facing humanity, even though scientists had been warning about it since the nineteenth century.

So what exactly is the greenhouse effect? Paradoxically, it is essential for life on earth. The sun heats the surface of the earth and this heat is radiated back into space by the planet. Some of the heat from the earth is absorbed by gases in the atmosphere, rather like a greenhouse trapping heat from the sun. This results in the earth maintaining an average temperature of +15°C. If there were no greenhouse effect and the heat was simply radiated back to space then the average temperature would be −18°C and life as we know it would not have evolved on earth.

In the last few centuries the relative concentrations of gases in the atmosphere have changed as a result of man's activities and there are now more heat absorbing gases in the atmosphere than there were. The result is that the earth's average temperature is set to rise slowly. In fact global temperatures have been increasing over the last few decades (by 0.5°C since 1900) and some scientists believe the greenhouse effect is the cause. Even a small increase could have highly significant consequences. At the height of the last ice age 18 000 years ago the average temperature was just 4°C lower than it is today. Yet estimates suggest that by the years 2030–2050 temperatures could have risen by between 1.5 and 5.5°C. That represents a climate change beyond any experienced in the last 10 000 years. Changes in climate are nothing new. Measured on a scale of millions of years the world's climate has been remarkably unstable. In the last 1.5 million years there have been 17 periods when the climate in mid-latitudes has been significantly colder – by about 9°C. Now there are fears that mankind may be precipitating another change – but in the opposite direction.

The suggestion that burning coal might upset the delicate balance of gases in the atmosphere was first proposed nearly a hundred years ago in 1896 by a Swedish chemist Svante Arrhenius. He suggested that the Industrial Revolution would release so much carbon dioxide into the atmosphere that global temperatures would rise. In the 1930s there was renewed concern among scientists following a noticeable warming in the Northern hemisphere which had begun in the 1920s. For the first time the possibility of a man-made greenhouse effect began to make an impact outside scientific circles. Interest waned, however, as the middle latitudes in the Northern hemisphere cooled in subsequent decades. Elsewhere in the world temperatures continued to rise, though such information was not readily available to scientists at the time. In 1957 the Scripps Institute of Oceanography suggested that man was 'engaged in a great geophysical experiment'. Charles Keeling, a graduate student from the Institute, set up a measuring station on the volcano of Mauna Loa to test the relatively clean air of the mid-Pacific. Since 1958 he has measured an 11 per cent increase in the concentration of carbon dioxide.

CARBON DIOXIDE

Carbon dioxide, which is the main greenhouse gas, makes up 0.03 per cent of our atmosphere – a share which has varied by only 40 per cent over the last few million years. Because the variation has been so relatively slight it has allowed a life-sustaining climate to develop on earth. Other planets have not been so lucky. Venus, where the atmosphere is mostly carbon dioxide, suffers a runaway greenhouse effect producing scorching temperatures in which life cannot survive.

TABLE 1 EMISSIONS OF CARBON WORLD WIDE FROM MAN-MADE SOURCES

(millions of tonnes a year)	
1950	1639
1955	2050
1960	2586
1965	3154
1970	4090
1975	4628
1980	5249
1985	5338
1986	5555

Source: *Environmental Data Report 1989/90* (United Nations Environment Programme)

CARBON EMISSIONS FROM FOSSIL FUELS 1960 AND 1987

	Carbon (million tons)		*Carbon per head of population (tons)*	
	1960	*1987*	*1960*	*1987*
United States	791	1224	4.38	5.03
Soviet Union	396	1035	1.85	3.68
China	215	594	0.33	0.56
United Kingdom	161	156	3.05	2.73
West Germany	149	182	2.68	2.98
France	75	95	1.64	1.70
Japan	64	251	0.69	2.12
Poland	55	128	1.86	3.38
Canada	52	110	2.89	4.24
India	33	151	0.08	0.19
Australia	24	65	2.33	4.00

Source: *State of the World 1990, A Worldwatch Institute Report on Progress Toward a Sustainable Society* (W. W. Norton & Co) © Worldwatch Inst.

Before the Industrial Revolution, there were 280 parts per million of carbon dioxide in the atmosphere. But the Industrial Revolution was based on the burning of coal and, together with the burning of other fossil fuels like oil and gas, man has in the last hundred years or so been rapidly releasing the carbon which has been locked up over the previous 300 million years. At the present time there are about 350 parts per million of carbon dioxide in the atmosphere and this figure is increasing by 1.5 parts per million each year.

By measuring carbon dioxide concentrations in air bubbles trapped in ice it is possible to find out what the concentration of the gas was in the atmosphere thousands of years ago. The grim fact is that it is now at its highest level for 160 000 years.

In the late 1980s the combustion of fossil fuels released about 5.5 billion tonnes of carbon a year into the atmosphere – that is about a tonne for each human being on earth. Industrialised countries tend to produce more, developing countries far less per head of population. In some Eastern bloc countries like the Soviet Union, where market incentives do not have any great effect on industries, there have been no pressures to improve energy efficiency and while the economic output per head of population is only two-thirds that of Western Europe, carbon emissions per head are twice as high.

There is no sign of a slowing down in the use of fossil fuels – just the opposite. Fossil fuels provide nearly four-fifths of the world's energy and their use is projected to grow at the rate of 1–2 per cent a year for the foreseeable future. For example, if Britain continues to increase its use of energy at the current rate, by the year 2005 it will have increased its carbon dioxide emissions by 37 per cent and by the year 2020 by 73 per cent.

Even if the rate at which carbon dioxide is emitted world wide remains constant from now on, the level of carbon dioxide in the atmosphere will have doubled in 150 years. But if the rate at which carbon dioxide is being put into the atmosphere increases, as many think it will, then doubling could occur earlier – in 80 to 130 years. And there's no reason to think once it has doubled the increase would stop there. There are sufficient fossil fuel reserves to increase the atmospheric concentration of carbon dioxide some five to ten times.

The burning of fossil fuels is not the only problem. The burning of many of the world's forests is also aggravating the situation contributing up to 20 per cent of carbon dioxide in the atmosphere. Not only does the burning itself release carbon dioxide but the effect is compounded because with huge acreages of forests being removed there is that much less photosynthesis occurring and therefore less carbon dioxide being removed from the atmosphere. By 1980 about 11 000 square kilometres of forests were being cleared annually. It is estimated that over recent decades between 800 million and 1.6 billion tonnes of carbon (as carbon dioxide) has been released every year into the atmosphere from this source. Today if deforestation were stopped perhaps up to 2 billion tonnes of carbon a year would be prevented from going into the atmosphere. Brazil alone contributes an estimated 336 million tonnes of carbon to the atmosphere each year by burning its forests – six times as much as through burning fossil fuels. That makes Brazil the fourth largest emitter of carbon into the atmosphere in the world.

There is a fixed amount of carbon in the world. Most of it is locked into rocks and sediments and plays no part in the greenhouse effect. An estimated 12 000 billion tonnes of carbon is contained in fossil fuels and shales of which about 7500 billion tonnes are ultimately recoverable. In addition the oceans contain another 38 400 billion tonnes, plants and animals contain another 1760 billion tonnes and in the atmosphere there is a comparatively small 743 billion tonnes.

There is a constant flow of carbon between these various reservoirs. Each year an estimated 100 billion tonnes of carbon is given up by the oceans to the atmosphere and a similar amount is absorbed by the oceans. Equally each year another 100 billion tonnes is given out by the plants and animals on land and a similar amount is absorbed by growing plants. So long as the flow remains in balance then the climate will remain stable. But man's activities have changed the balance. Of the 5.5 billion tonnes being put into the atmosphere every year by man's activities about half is absorbed by the oceans – the rest stays in the atmosphere to build up year by year and increase the greenhouse effect.

Scientists accept that there are still a lot of uncertainties surrounding what happens to carbon dioxide in the atmosphere. As we have said, much is absorbed into the oceans. But some is removed in other ways. Raised levels of carbon dioxide might be increasing plant growth by encouraging photosynthesis – a sort of fertiliser effect – but, it's thought, this is probably removing very little. Some of the excess carbon dioxide might be absorbed by the increasing growths of algae in the oceans which have been stimulated by the excess nutrients flowing into the sea from rivers polluted with agricultural fertiliser. And several other possible ways have been suggested by which carbon dioxide might be removed from the atmosphere under natural conditions. But even if they were real effects the amount of gas removed would be very little.

Carbon dioxide is not the only greenhouse gas. Others include methane, CFCs, nitrous oxide, ozone and water vapour. From 1850 to about 1960 the increase in the carbon dioxide concentration was the dominant factor in encouraging the greenhouse effect. But since then the combined effect of increases in all the other gases has been equal to that of the increase in carbon dioxide.

METHANE

The concentration of methane in the atmosphere was roughly stable until about 200 to 500 years ago when it began to grow. It is now present in the atmosphere at a concentation of 1.7 parts per million – about double what it was in the Middle Ages, and it is increasing by one and a half per cent a year.

Methane comes from rice paddies, termites and other animals, waste disposal and oil recovery. On average 540 million tonnes of methane is released into the atmosphere each year – a lot of it from natural sources. The most important source of methane in the UK is leakage from landfill waste sites. British Gas believes methane from waste sites is equivalent to 7 per cent of the gas they supply each year.

Animals contribute a considerable amount of methane to the atmosphere. A cow, for example, produces 200 grammes per day through flatulence. With an estimated 1300 million cows in

TABLE 2 AMOUNT OF METHANE IN THE ATMOSPHERE

Year	*Methane (parts per billion by volume)*
1654	650
1771	780
1804	730
1861	830
1919	1000
1950	1180
1955	1300
1965	1388
1975	1450
1980	1590
1984	1626

Based on direct measurements in the atmosphere and measurements of methane trapped in polar ice.

Source: *Environmental Data Report 1989/90* (United Nations Environment Programme).

the world that makes 100 million tonnes of methane going into the atmosphere each year from cows alone!

Methane is also increasing in the atmosphere because the natural processes which can remove methane have been impaired by the emission of other man-made gases like carbon monoxide. Between 10 and 40 per cent of the rise in methane may be due to a reduction in the natural rate of its decay.

Millions of tonnes of methane are trapped in the frozen tundra. Should the tundra melt with rising temperatures much of this methane will be released, increasing the greenhouse effect still further. Methane is 25 times more efficient as a greenhouse gas than carbon dioxide and contributes about 12 per cent to the current greenhouse warming.

CFCs

CFCs are very stable gases which have been used for a variety of applications. They were used extensively as propellants in aerosols, in foam blowing, as refrigerants and as cleaning agents in the electronics industry. Concern over what CFCs are doing to the ozone layer has led to international agreements to limit their use. Now in Britain they are used in less than ten per cent of aerosols, their use in refrigerators and foam blowing is being reduced, and there is pressure for their use to be phased out as cleaning fluids (see Chapter 5). Molecules of CFCs can last as long as 100 years in the atmosphere and it will be many years before the present reduction in their use will cause any marked difference in their concentrations in the atmosphere. At the moment global CFC levels are still increasing by between 4 and 6 per cent a year.

CFCs are said to be 10 000 times more potent as greenhouse gases than carbon dioxide because they absorb infrared radiation in a region of the spectrum where there is little absorption by other gases. Although they are present in the atmosphere in only very small amounts – just 0.000225 per cent of the atmosphere – they contribute some 25 per cent to the current greenhouse warming. The use of CFCs in the United States alone contributes 40 per cent of that country's contribution to global warming.

NITROUS OXIDE

Nitrous oxide is produced mainly by the action of bacteria, from fertilisers and through deforestation. It is present in the atmosphere at a concentration of 0.28 parts per million and is increasing at the rate of 0.3 per cent a year. Nitrous oxide molecules are long lasting – their lifespan is estimated at about 150 years. Molecule for molecule nitrous oxide is 250 times more efficient as a greenhouse gas than carbon dioxide. Currently nitrous oxide is responsible for about 6 per cent of greenhouse warming.

OZONE

Ozone is also a greenhouse gas. The ozone is formed at ground level, partly as a result of photochemical reactions between hydrocarbons (like methane and petrol vapour), and nitrogen oxides. But ozone is also being depleted in the upper atmosphere due mainly to CFCs. It is believed by some scientists that the two trends will cancel each other out, leading to a negligible effect on global warming.

WATER VAPOUR

Water vapour is a greenhouse gas but exactly what effect increasing amounts of water vapour will have on the atmosphere is uncertain. As the earth heats up more water evaporates and therefore more water vapour is produced. The resulting increased cloud cover, however, could have two opposing effects. Some clouds would reflect more of the sun's heat back into space so tending to reduce global warming. On the other hand, by acting as a blanket over the land, other types of clouds tend to trap some of the heat radiated from earth. The net effect depends on the relative change in the different cloud types and their height and has yet to be ascertained. Apart from being a greenhouse gas, water vapour in the atmosphere helps speed up chemical reactions there. This helps the removal of methane and other hydrocarbons from the troposphere. These complex chemical reactions also produce and destroy another greenhouse gas – ozone. The net effect could well be that increasing water vapour would actually decrease the greenhouse effect.

HAS THE GREENHOUSE EFFECT STARTED?

According to a report on the greenhouse effect from the Royal Society published in 1989, the changes in the global and regional climates due to greenhouse gases 'will be small, slow and difficult to detect above natural fluctuations during the next 10 to 20 years.' Natural fluctuations on this time scale can be caused by a variety of things including variations in the sun's output, volcanoes, and changes in the ocean's currents. Detecting any greenhouse effect will be difficult when one considers the huge differences in temperatures from one day to the next and from one part of the world to another.

But some experts believe the beginnings of the greenhouse

effect are now detectable and appreciable changes will occur within the next ten years.

One scientist who believes there is clear evidence that the greenhouse effect has begun is James Hansen – the director of the NASA Goddard Institute of Space Studies. In evidence to a US Senate hearing in 1988 he produced figures which he said indicated the greenhouse effect was now upon us. He stressed that global average temperatures going back to 1880 showed an increase of 0.6°C. And he pointed out that the decade of the 1980s had been the warmest ever recorded. The six warmest years on record have been, in order, 1988, 1987, 1983, 1981, 1980 and 1986. His testimony caused a considerable flutter in the dovecots. Whilst many scientists admit that there is a strong possibility that the observed warming is due to the greenhouse effect few are prepared to state categorically that it is the sole cause. Uncertainties will only be resolved if and when further warming surpasses the natural variability in climate. One leading scientist, Stanley Grotch of the Lawrence Livermore Laboratory, believes that if there were a secret ballot on the question most scientists would agree that the man-made greenhouse effect is the most plausible explanation for the warming which has already occurred. Whatever the scientific justification for Hansen's statements he did succeed in making the greenhouse effect the focus of public attention and that certainly sharpened the minds of politicians.

MEASURING GLOBAL WARMING

Calculating the effect on the climate of increases in carbon dioxide in the atmosphere is no simple matter. Powerful computers are provided with a set of assumptions and then given a certain amount of raw data to work on. They then perform millions of calculations and produce an estimate of future global warming. These climate models are really elaborations, on longer time scales, of course, of the models used to predict our weather. The models are being refined all the time but a measure of their uncertainty is shown by the fact that a slight adjustment of the model can have a major effect on the result produced. For example, the Meteorological Office in Bracknell, Berkshire, which has one of the most advanced models of the global atmosphere in the world, recently predicted that a doubling of carbon dioxide in the atmosphere could produce a temperature increase of 5.2°C, together with a 15 per cent increase in precipitation (rain, hail and snow). The model also revealed that the temperature rise would be much higher in the polar regions and that rainfall would increase most in middle and high latitudes. Recently, however, the Met Office has modified the model to take into account the fact that some of the water present in clouds is in the form of ice crystals which have different optical properties from water. Altering the data in the model in this way resulted in a predicted temperature rise of not 5.2 but 2.6°C – a clear demonstration of how fragile some of the modelling is. Indeed, meteorologists studying the climate now believe that gathering better information about clouds (their position, height and type, as well as the optical properties of the water they contain) is one of the priorities of research.

To decide whether climate is changing two types of measurements are crucial – temperature and rainfall. It is difficult to compare measurements now with similar measurements from the past partly because the accuracy of past records may be open to question and also because record keepers then may well have used different techniques. When scientists are looking for a relatively tiny change in temperatures and rainfall over many decades accuracy is vital. To improve the modelling of the climate scientists need more powerful computers and also better information to feed into them. Satellite measurements go some way to filling in the gaps but existing techniques are imperfect and better ones are being developed.

The role of the oceans, too, is vital to our understanding of the greenhouse effect. Three-quarters of all the solar energy received by the earth is absorbed by the oceans and half the carbon dioxide emitted by burning fossil fuels dissolves into the oceans. Models are being improved and will benefit considerably from the vast amount of new information about the oceans which should result from the new oceanographic satellites due in the early 1990s.

Also of key importance in assessing the progress of global warming and its likely consequences is the state of the ice at the polar regions.

The Scott Polar Research Institute, with the use of nuclear submarines, has provided evidence that warmer seas are already having a noticeable effect on the thickness of the ice sheets and the sea ice. What is most worrying is that melting sea ice in the Arctic and Antarctic could have a positive feedback effect, because as the ice melts there is less of it to reflect the heat back into space and so warming is increased.

EFFECTS OF GLOBAL WARMING

Perhaps the most dramatic result of any substantial rise in global temperature would be a rise in sea level. Estimates vary but a reasonable guess is that over the next forty years (to 2030) the rise in sea level will be between 5 and 60 cm – with 20 cm the preferred estimate. About half of this rise would be due to the expansion of the seas as they heat up. The other half would be due to the melting of the ice in glaciers and ice present on land. (Melting sea ice would not affect sea level.) A rise of 1 metre before the end of the next century (2100) would affect up to 300 million people. The cost of protecting life and investment might exceed £13 billion a year. A more extreme calculation is that sea level could rise by between 1 and 2 metres in the next century which would threaten many coastal cities. Some would even be destroyed, especially in Third World countries which might not be able to afford to build extensive defences.

In some developing countries a sea level rise would result in salt water encroaching on vital rice paddies making them unusable. In Bangladesh, for example, it is estimated that by 2050, 18 per cent of the land mass could disappear beneath the waves displacing over 17 million people. Such a change might come about very quickly with tropical storms leading to huge areas being suddenly overwhelmed. Millions living in the delta areas of the Nile, the Ganges and the Yangtze would be forced out of their homes. A rise of two metres would mean the

Maldive Islands would simply disappear as the highest point on the islands is not more than two metres above sea level. And it is not just the Third World which is threatened. In New York the island of Manhattan is just four feet above sea level. Billions of dollars would have to be spent on defences and on maintaining the water supply.

One of the starkest warnings about sea level rise has come from Britain's ambassador to the United Nations Sir Crispin Tickell. In a lecture in June 1989 to the Natural Environment Research Council he warned that by the middle of the next century there could be as many as 300 million environmental refugees. 'The last period of warming', he said, 'showed a human invasion into the areas liberated by the ice . . . But in a new and more drastic period of warming there would be few places for people to go, for other people are there already. We have left ourselves no room for manoeuvre.'

Clearly there is a lot of uncertainty in forecasting the extent of the future rise in sea level. Calculations are complicated by the fact that while warmer temperatures will tend to melt the ice they would also tend to increase snowfall at the poles. This could simply replace the water on top of the ice sheets of Greenland and Antarctica.

Internationally, the effects of a significant global rise in temperature could be catastrophic for agriculture. Models suggest that the major grain growing areas of the United States and China will become substantially hotter and drier. The US corn belt could shrink to a third of its present size. Some people believe that the hot dry summer which the American mid-West experienced in 1988 may well be a foretaste of what is in store when the greenhouse effect bites. There could be compensating effects, though, with Canada and Siberia becoming much more productive as they heat up.

In Britain the conditions might encourage the growing of such crops as sunflowers and maize and possibly even tobacco and soybean. On the other hand there would probably be less oil seed rape grown. If Britain became drier as a result of climate change it might shift the main areas of arable farming away from the Eastern Counties towards the West. It could effect our economy too: if the rest of Europe became very much drier than Britain then we might find we would become the grain bowl of Europe.

It is possible new strains of crops could well be developed to tolerate the new climatic conditions. As the recent report on the greenhouse effect from the Royal Society pointed out, 'it was economics and the plant breeder that pushed the crop into different climates rather than climate change that moved the crops.'

PREVENTING THE CATASTROPHE

If energy strategies have to change drastically to limit the emissions of carbon dioxide what can they change to? One suggestion is that a switch from coal and oil burning to gas burning could reduce emissions. Coal produces 75 per cent more carbon per unit of energy, and oil 44 per cent more. But though there is plenty of natural gas globally it is ultimately a finite resource. And producing more gas would increase the leakage of methane from natural gas distribution systems. It is said that this leakage would itself reduce any benefit to the greenhouse effect achieved by burning the gas.

What of the option of removing the carbon dioxide from flue gases? That would be expensive and impractical. Carbon makes up about 73 per cent of coal by weight. Huge amounts would have to be extracted and disposed of – seven times the amount of ash which now has to be disposed of from coal-fired power stations. And then how would this carbon be kept from the atmosphere? It has been suggested that it could be buried deep at sea – but that would be expensive and require long pipelines. It could also have unforeseen consequences.

Perhaps the best and most realistic way of cutting greenhouse gases is to use energy more efficiently. Improving energy efficiency might cut fossil consumption by 2 per cent a year in industrialised countries. Improving the efficiency of appliances which use electricity like refrigerators and light bulbs could reduce the amount of fossil fuel burnt. Lighting accounts for 17 per cent of electricity use – producing 250 millions tons of carbon emissions a year. More efficient light bulbs could reduce that by half.

It would be possible to build combined heat and power plants to provide electricity to apartment blocks or even whole cities – again leading to a significant saving. Changing the way energy is generated as a way of combatting the greenhouse effect is considered in more detail in the next chapter.

But fossil fuels are not burnt just in power stations. The world contains nearly 400 million cars. They emit 550 million tons of carbon each year into the atmosphere – 10 per cent of the total from fossil fuels. If all cars were made so that they did 50 miles to the gallon rather than the current average of 20–30, emissions would decline.

Britain's carbon dioxide emissions in the domestic and commercial sector have remained fairly constant since 1977, partly as a result of the major change in Britain's industrial base. The main growth has been in emissions from traffic which made up 16 per cent of the total in 1987 compared with 11 per cent ten years earlier.

Another way of countering the greenhouse effect would be to plant more trees. Trees use carbon dioxide in photosynthesis. World wide, growing plants soak up some 120 billion tons of carbon a year – more than 20 times the amount of carbon released by burning fossil fuels. It has been calculated that planting an area of trees twice the size of France would absorb 660 million tons of carbon each year for the next three decades until the trees became mature – about ten per cent of net carbon emissions. But tree planting on this scale would be extraordinarily difficult to put into practice and would not be cheap especially if the land on which the trees are grown had to be purchased.

There is not a great deal Britain alone can do by way of reducing the greenhouse effect by reforestation, although it has been estimated that if the amount of land surface in Britain covered by trees were doubled from the present 10 per cent to 20 per cent, and broad-leaved trees planted, that would absorb some 3 million tons of carbon each year.

Global warming could also be slowed by reducing deforestation. Halving the rate in Brazil, Indonesia, Colombia and Côte d'Ivoire could reduce net carbon emissions from tropical forests by more than 20 per cent.

Even if countries did reduce the output of carbon dioxide in the future there is still the prospect that sea levels will rise as a result of warming which has already been triggered. *So how might the problem of rising sea levels be coped with?* Combatting the expected rise in sea levels would involve massive expenditures to reinforce sea defences. But scientists have come up with other suggestions to help slow it down. One bizarre idea is to pump the extra water into low-lying depressions in the earth's surface making in effect a huge inland sea with the surplus water. The Caspian Sea lies in just such a large depression. By building a pipeline from the Black Sea to the Caspian Sea it might be possible to pump the extra water into the Caspian Sea enlarging it over a period of some twenty years. Those suggesting this idea do not believe it is in any way an answer to the greenhouse effect, but it might slow one of the more dramatic consequences.

One of the great 'sinks' for carbon dioxide in the oceans is the plant life. It is estimated that some ten billion tons of carbon dioxide are 'fixed' each year as a result of photosynthesis by plant life in the oceans. Seaweeds use some 10 per cent of it and the rest is used by the various types of plankton. Scientists have been interested in what would happen to this plankton if there were a rise in sea temperature or an increase in carbon dioxide in the oceans.

Of the carbon dioxide which disappears into the oceans it had been thought that about half dissolves into the water and the other half is taken up by the plankton. In fact, research results from a 15-month expedition in the North Sea by the British Research Ship *Challenger* has shown that only about 30 per cent of the carbon dioxide dissolves into the water. The plankton, therefore, must be playing a more important role than scientists had thought.

It follows that if plankton could be encouraged to grow in areas where they are fairly sparse at the moment that might pull much more carbon dioxide out of the air. Research in the polar regions shows that the factor limiting the growth of plankton there is the absence of a trace element – iron. If soluble iron could somehow be put into the southern ocean then huge new areas of plankton could be encouraged. The Natural Environment Research Council is due to carry out an experiment in the Southern Ocean to test the theory to see if it could modify the greenhouse effect in any way.

There are all sorts of feedback mechanisms which operate in the atmosphere and the plan to boost photosynthesis in the Southern Ocean may face more difficulty than imagined. Recent research has shown that there is increased ultraviolet radiation over the polar region as a result of CFCs in the upper atmosphere destroying the ozone layer. The ultraviolet radiation reduces the rate at which plankton photosynthesise with the result that they are probably taking up less carbon dioxide than they once were.

The full range of measures needed to prevent the greenhouse effect getting any worse were neatly summed up by the US Environmental Protection Agency. In a report to the American Congress, it stated that a reduction in the greenhouse effect will involve: price increases for coal and oil, many more solar power devices, more use of nuclear and biomass energy, new forests, big reductions in the use of chlorofluorocarbons and related products, new ways of producing rice, meat and milk, systems in landfill sites to capture methane gas and an unusual degree of cooperation among the developed and the developing nations. The journal *Science* reported: 'if all these things begin happening in the early 1990s, the rate of gas build up may level off in the 22nd Century'.

The relatively low oil prices in the 1980s have already led to a slowing down of energy conservation measures and efforts to improve energy efficiency. The worldwide slow down in the development of nuclear energy has also meant the prospects for a reduction in carbon dioxide emissions are not particularly good.

A Worldwatch Institute report in October 1989 estimated that industrialised countries are responsible for 46 per cent of the problem, the Soviet Union and Eastern Europe 19 per cent and the developing world, 35 per cent. The report stresses what many developing countries themselves have been saying – that the industrialised world, having caused most of the problem, has a responsibility to lead the way in finding, and paying for, solutions. Estimates vary, however, as to how drastic the cutback in emissions would need to be to stabilise the climate.

HOW COMMITTED ARE THE POLITICIANS?

Increased recognition of the seriousness of the problem has led to many international conferences in the last year or so at which scientists and politicians have discussed the issues in depth. In June 1988 an international conference of scientists in Toronto, attended by the Prime Ministers of Canada and Norway, suggested a goal for the year 2005 of a reduction in carbon dioxide emissions of 20 per cent of 1988 levels and of 50 per cent eventually in order to stabilise the atmosphere. Their initial target would certainly slow down greenhouse warming but it would not stop it. A detailed report on how a 20 per cent reduction might be achieved by 2005 was produced in 1989 as a result of research funded by the Dutch Government. It suggested that the amount of carbon dioxide which could be allowed to be released between 1985 and 2100 was 300 billion tons. The budget was based on a calculation of the rate at which carbon can be allowed to continue to increase if average rises in temperature are to be kept to 0.1°C per decade. The budget was also based on a ceiling for temperature rise of 2.5°C above that of the average temperature at the start of the Industrial Revolution. At current rates of release that 'budget' would be used up by the year 2030. The report suggested that the budget should be divided equally between industrialised and developing countries. This would mean that industrialised countries would have to implement fairly drastic cuts to meet the budget while developing countries could actually increase their carbon emissions for a while. It suggested that national budgets would be based on rates of release of carbon dioxide per head of population. On this basis Britain would be allowed to continue emitting carbon at current rates for 42 years before using up its budget. Thereafter it would have to start reducing output.

In November 1988 representatives of some 30 nations met in Geneva under the auspices of the United Nations Environment Programme and the World Meteorological Organisation to review the science and policy options relating to climate change, and formed an Intergovernmental Panel on Climate Change to forge an agreement.

Three working groups were set up. One, headed by Britain, was charged with the job of reviewing the scientific evidence. A working group headed by the Soviet Union was to look at the possible effects of climate change and the third group headed by the United States was to look at what strategies might be adopted to slow down global warming as well as how to adapt to it.

It was intended there should be a draft action plan in time for the Second World Climate Conference in 1990.

Meanwhile the enthusiasm of the politicians gained ground. In December 1988 the United Nations agreed a special resolution calling for the adoption of a framework convention on climate change. And the European Commission issued a report calling for a comprehensive continent-wide effort to analyse and cope with global warming.

In March 1989 in The Hague the leaders of seventeen countries urged that a strong new international institution with enforcement powers to carry out the provisions of a global warming agreement might be needed – a sort of United Nations Environment Programme with teeth. And at the Paris economic summit in July 1989 there was agreement on the need to cut carbon emissions.

The most important intergovernmental meeting on the greenhouse effect so far took place in November 1989 in Noordwijk and was attended by 69 nations including the United States, the Soviet Union, Japan and Britain. It agreed to a freeze on carbon dioxide emissions 'as soon as possible'. This form of words was substituted for a target date of 2000 because some countries felt it would be unrealistic to set a specific date until the International Panel on Climate Change had reported in 1990. Some industrialised countries at the conference went somewhat further and committed themselves to stabilising carbon dioxide emissions from power stations and motor vehicles by the year 2000 'at the latest'. Although the agreements were fiercely criticised by environmental organisations for being too vague, for Britain at least any such agreement would have profound implications. With motor traffic increasing dramatically it would seem a well nigh impossible task to guarantee that carbon dioxide emissions could be stabilised even with ten years to do it in. It either implies a major expansion of public transport, a major technical breakthrough to control emissions or some more profound method of limiting growth in motor vehicles. The last two options seem to be particularly remote and there has been little enthusiasm by Mrs Thatcher's Government for the first. The agreement to stabilise carbon dioxide emissions from power stations implies a major expansion of energy conservation measures. British estimates suggest that on current trends carbon dioxide output is likely to have increased by 20 per cent by the year 2000.

The conference also reached agreement on the need for a world treaty to protect the atmosphere. And there was agreement on exploring the possibility of setting up a world atmosphere fund to help developing countries change their energy policies, and a target of planting 29 million acres of trees a year by the year 2000 was set. Environmentalists were somewhat dismissive of the outcome of the conference, claiming that the agreement would not save one tonne of carbon dioxide going into the atmosphere – but it was a landmark conference which set important goals. The day after the agreement was reached Mrs Thatcher made the environment the main issue when she addressed the United Nations in New York. She announced that Britain was setting up a new Centre for the Prediction of Climate Change, thus making more specific, and implying expansion of, work which had already been going on at the Meteorological Office.

Pressure for action to reduce the output of greenhouse gases boiled over at a preliminary 'housekeeping' meeting of the Intergovernmental Panel on Climate Change held in Washington in February 1990. Britain was one of several countries who favoured waiting until the various groups preparing scientific and technical reports had presented them to the second meeting of the Panel in Geneva in September. But a group of nine countries – Austria, France, Denmark, West Germany, Italy, Sweden, the Netherlands, Norway and Switzerland – said they were to begin work on their own treaty to limit greenhouse gases. They said they were concerned at any delay in action which might endanger future generations.

Despite the increased determination by some politicians that major steps need to be taken, not all countries are equally persuaded. The American Government has begun to show its willingness to take the matter seriously though some have felt it has been somewhat hesitant in its action. Its first official statement on the issue, in January 1989, came from Secretary of State James Baker who said 'we can probably not afford to wait until all of the uncertainties have been resolved before we do act'. But his priorities focused on actions that could be taken which were already justified on grounds other than their threat to the climate, including reducing CFC emissions, greater energy efficiency and reforestation.

If American enthusiasm was only lukewarm in January 1989, by November it had cooled still further mainly as a result of a thirty-five-page report by three distinguished scientists – a former Director of the Scripps Institute of Oceanography, the founder and former director of the Goddard Institute for Space Studies and a past president of the National Academy of Sciences. Their report said that greenhouse predictions were full of uncertainties and it was too soon to take any action. They said there was no evidence that the half a degree centigrade rise in temperature that had occurred this century was caused by greenhouse gases. It was more likely to have been caused by increased solar activity. And they predicted that decreased solar activity in the next century would lead to a cooling trend which would probably offset any rise in temperature caused by greenhouse gases. Their report stirred up a hornet's nest of opposition from many equally distinguished scientists who said the report was biased, misleading and 'unrefereed'. One said it was based on 'junk science'. It had its supporters though. One, Richard Lindzen, Sloan Professor of Meteorology at the

Massachusetts Institute of Technology, says he believes the computer models that predict a greenhouse warming are fatally flawed because they don't take account of some of the negative feedback mechanisms by which a warmer atmosphere can re-adjust itself – claims dismissed by Stephen Schneider of the National Centre for Atmosphere Research at Boulder, Colorado who retorts: 'does he have a calculation or is his brain better than our models?'

Some countries are actively considering plans to control carbon emissions but in general, despite the rhetoric, there has been little firm action so far. As the Worldwatch Institute says, 'If the international community cannot get beyond the level of grand rhetoric by the 20th anniversary of the Stockholm Conference in 1992, it will fast be getting too late to stabilise the climate.'

In Britain the Prime Minister would appear to be taking the issue seriously. She held a special seminar for her cabinet ministers in April 1989 which was addressed by many leading scientists. Her Government spent £15.54 million on research and development projects on global warming in 1989–90 as well as contributing some £760 000 towards the Intergovernmental Panel on Climate Change. But out of a total Government research budget of £5500 million it is perhaps not conclusive proof of the Government's recognition of the magnitude of the problem.

If Britain decided to implement the target set at the Toronto conference of a cut in carbon dioxide emission of 20 per cent by the year 2005 how could it do it?

According to the Central Electricity Generating Board (now split up into a number of smaller companies) there are several scenarios. Simply switching methods of electricity generating to cut carbon dioxide emissions would mean closing down nine coal-fired power stations and replacing that electricity with **more oil and gas plants** and more nuclear capacity. It would require an investment of £12 billion and the effect on electricity prices would be 'very severe'.

Switching to nuclear power stations would probably reduce carbon dioxide emissions the most but it would be one of the most expensive options and there would be considerable public opposition.

The CEGB also looked at how to achieve the same goal by maintaining current projected generating methods and reducing demand for energy by 37 per cent by **energy efficiency measures**. That would mean that by 2005 energy demand would still be at the present level and would have resulted in a big reduction in the burning of coal. It would, however, 'require Draconian intervention in the energy and equipment markets and considerable social change'. The privatisation of the electricity industry has already resulted in a market-led change in electricity supply with increasing emphasis on burning gas and the projected closure of a number of coal-burning power stations. This will have a small but favourable impact on the emissions of carbon dioxide.

In June 1990 the IPCC panel for which Britain was responsible reported its conclusions – and they were stark. It said that to stabilise carbon dioxide at today's levels it would require a 60 per cent cut in emissions of long-lived greenhouse gases such as carbon dioxide and CFCs and oxides of nitrogen. Emissions of the shorter lived greenhouse gas, methane, would need to be cut by 15–20 per cent.

It warned: 'the longer emissions continue to increase at present-day rates, the greater reductions would have to be for concentrations to stabilise at a given level'. It made the point that unless cuts in emissions were made the earth would heat up by an average of 3°C before the end of the next century and sea level would rise by 20 cm by 2030 and by 65 cm by 2100.

The specific warnings brought a swift response from the British Government which said that it would aim to reduce by 30 per cent the presently projected levels of carbon dioxide emissions by the year 2005. It was a target which Mrs Thatcher said would be very demanding but would be implemented 'provided others are ready to take their full share'. Predictably some environmentalists complained the British pledge did not go nearly far enough.

Perhaps one of the most sensible and realistic ways of reducing carbon emissions would be for governments to institute a

carbon tax. Those fuels which produce the most carbon, like coal, would be taxed highest. Oil would be taxed somewhat less, natural gas, which produces the least carbon per unit of energy, would be taxed least and of course renewables which don't produce any carbon would carry no tax at all. That would give considerable financial incentive to both domestic and industrial users to opt for the least polluting sources of energy. The tax would put up the cost of energy – but by nothing like as much as the cost of having to deal with the greenhouse effect.

The Worldwatch Institute estimates that a tax of, say, $50 a ton of carbon levied world wide would bring in about $280 billion a year. In countries which use a lot of energy like the United States that would cost each individual an average $240 a year, while in India it would cost an individual just $9. Such a tax would raise the price of electricity by 28 per cent.

Clearly any international agreement to cut carbon dioxide output to the required levels would have a dramatic impact on the economies of the developed countries. One recent report by the Royal Institute for International Affairs has pointed out the difficulty of negotiating such an agreement: 'Trying to negotiate targets for each country makes the participants focus on why they shouldn't reduce as much as the next. The difficulties in reducing emissions are great enough without the diplomatic prizes going to those who can amplify them the most.' The report suggests the unusual idea of countries having permits for carbon emissions which are tradeable so that a country like Japan, for example, instead of having to cut its own emissions, might be able to reduce global carbon dioxide more by helping a developing country reduce its emissions by some form of technological aid.

But grand plans for reducing the greenhouse effect by increased energy efficiency, cutting back on the use of fossil fuels and planting huge areas of forests have to be paid for, and it is the developing countries where such projects will be most difficult to implement given that they already have such huge debts. One suggestion has been that a Global Atmosphere Fund could be established using perhaps just 10 per cent of any international carbon tax.

The World Bank and the UN Development Programme could also help. The World Bank could perhaps lend money for efficiency improvements. But no one should be in any doubt whatsoever about the magnitude of the task ahead. It has been calculated that if agreement was reached to aim to stabilise greenhouse gases by the middle of the next century, when the world's population will have reached 10 billion, carbon emissions per head of population would have to be no bigger than India's carbon emissions are today – in fact one-tenth of the current European level.

Because it is such a dauntingly big problem, one wonders if there would be any point in one country like Britain taking unilateral action ahead of international agreements. The answer has to be yes – for two reasons. Firstly, it would set an example to the rest of the world. And secondly – every little helps. To quote Edmund Burke: 'nobody made a greater mistake than he who did nothing because he himself could do only a little.'

CHAPTER THREE

Preventing the Greenhouse Effect

The greenhouse effect is the greatest environmental threat facing mankind and there are only a limited number of strategies that can be adopted to delay its onset and attempt to reverse the trend. One lesson, which is being recognised increasingly, is that the world uses too much energy of the wrong sort. The most obvious way of attempting to combat global warming is to cut down on our use of fossil fuels. For it is this major source of energy, with its release of carbon into the atmosphere, which is the single biggest cause of the greenhouse effect.

There are four possible ways in which the world might reduce the output of carbon when producing energy:

1 *It can try to use less energy.* It would be unreasonable to expect Third World countries, which have not yet achieved the standard of living enjoyed in the industrialised countries, to cut down their use of energy, so the burden would have to fall on the developed world. One-fifth of the world's population uses over 70 per cent of the world's commercial energy, so clearly that is where the energy savings might best be achieved. The problem is that industrialised countries depend on energy to maintain their wealth, so any substantial and enforced reduction would be counter-productive.

2 *The world could try to use energy more efficiently.* Current ways of using energy are very wasteful but it has been proved in the last decade or so that it is certainly possible to maintain a good standard of living without excessive energy use. Between 1973 and 1985 in the United Stated the gross national product grew by 40 per cent while energy consumption remained constant,

in effect, a decrease in the amount of energy used to produce a unit of gross national product of one-fifth.

It takes man just one year to burn the amount of fossil fuels which it took nature a million years to produce. The amount of energy used by man has gone up from the equivalent of burning 3570 millions barrels of oil in 1900 to burning 54 060 million barrels of oil by 1988. It is now accepted that every effort should be made not to waste energy – the next step must be a commitment to improving energy efficiency.

3 *The world can increase its use of gas over other fossil fuels.* Burning gas produces less carbon dioxide for the same amount of energy output than either coal or oil. For every one billion joules of energy produced, burning natural gas produces 14 kilogrammes of carbon dioxide, burning oil produces 20 kilogrammes and burning coal, 24 kilogrammes. There is, however, only about forty years' supply of natural gas left in the world's known reserves, most of it – some 70 per cent – controlled by the Middle East and the Soviet Union.

One way or another, then, the world is almost totally dependent on polluting, finite energy sources. Clearly new energy sources will have to be developed, which brings us to the last of the four options:

4 *We can adapt to non-carbon dioxide producing energy resources such as nuclear power and renewable sources of power* such as solar power and power from wind, waves and tides.

The next chapter will deal with nuclear power in more depth, so what of renewable sources of power? In the foreseeable future these are likely to produce only a relatively small fraction of the energy that will be needed, nevertheless they are in theory very attractive from various points of view and are worth looking at in more detail.

SOLAR POWER

Every 43 minutes the amount of sunlight falling on the earth is equivalent to the total amount of energy mankind uses in a whole year. The amount falling on the earth in one week is equivalent to all known coal reserves. But there are two disadvantages about power from the sun which make its exploitation by man difficult: it is intermittent and it is diffuse so that to trap it properly would require huge areas of land. Nevertheless there are several ways in which energy from the sun is being exploited and some of them show great promise for the future.

In nature the energy of the sun is stored by plants by the process known as photosynthesis. The sun's energy is used by the plant to build complex chemicals which act as an energy store. Ultimately all life depends on this process. Efficient photosynthesis has yet to be copied by man, but man can trap the heat from the sun and use it to cause a chemical reaction which produces a form of stored energy. This involves the use of so-called solar furnaces. Large dish-shaped mirrors are used to focus the sun's rays onto a target where they concentrate the sun's energy 10 000 times. In one experiment using a solar furnace a hydrocarbon like methane is mixed with steam or carbon dioxide at temperatures of about 1000°C in the presence of suitable catalysts. Heat is absorbed and a mixture of hydrogen and carbon monoxide is produced – a gas which is very similar to coal gas. The beauty of this system is that it can form a 'closed loop', in the jargon of the scientists. In other words, using the power of the sun the gas could be produced in a remote desert region and piped to where the energy is needed. Here the process could be reversed, thus releasing the energy as heat. The chemical products (methane and carbon dioxide, for example) could then be transported back to the desert plant where they could once more be converted into the useful hydrogen and carbon monoxide gas. This system is known as a 'chemical heat pipe'.

Another way to harness the sun's energy is to use arrays of mirrors to focus the sun's heat onto a central furnace to produce steam which can then drive a turbine. In California one such project is expected to produce electricity for 8 cents a kilowatt hour. (Electricity produced from coal costs about 3 cents a kilowatt hour.) Solar power could not produce base load electricity in this way but its uses are increasing all the time as its costs decrease.

Solar power can also be converted directly into electricity using what are called photovoltaic cells. Sunlight falling onto a piece of semiconducting crystalline silicon knocks electrons loose and these flow to produce an electric current. A solar cell which uses a lens to concentrate sunlight onto a chip of single crystal silicon can convert about 28.5 per cent of the sunlight falling on it into electricity. More advanced photovoltaic cells use gallium arsenide on top of the silicon and this increases efficiency to about 30 per cent.

In a recent development announced by Boeing, scientists have succeeded in making a solar cell for use in space which converts 31 per cent of the available light falling on it into electricity. It also performs well on earth where its efficiency apparently reaches 37 per cent.

There are various new developments in photovoltaics. Very thin films of semiconductors, far thinner than a human hair, can respond to light and make electricity. The material can be sprayed onto glass or can be electroplated. These solar cells are not very efficient but can be produced cheaply and can generate a small amount of electricity even in relatively dull weather. Until recently the best thin film cell using silicon converted sunlight into electricity with an efficiency of just 9 per cent. Then the company Arco Solar produced a solar cell using copper-indium-diselenide (CIS) which reached an efficiency of 11.2 per cent – a major step forward. At the moment these solar cells cost about $100 a square metre to make. Ultimately it is thought the cost could be halved and if the efficiency can be boosted to about 15 per cent then these solar cells could become competitive with existing methods of generating electricity.

At present the applications of solar power are limited. Solar powered calculators, for example, use a thin film of a semiconducting material which converts sunlight into electricity with only 3 per cent efficiency. Such devices are also used in watches. Satellites and units requiring small amounts of energy in remote areas like deserts also make good use of photovoltaics. Markets for solar energy continue to expand as the price of solar power continues to drop.

Photovoltaic cells which convert sunlight directly into electricity now cost 30 to 40 cents per kilowatt hour. This would have to come down to 12 cents per kilowatt hour before it could be thought of as remotely competitive for peak load electricity generation, and down to about 6 cents if it is to be used for widespread generation of electricity.

In Britain there is considerable interest in incorporating photovoltaic panels into roofs of buildings to generate electricity for use in the building. One company, Chronar Ltd, is due to fit amorphous silicon solar panels to such a building in a £200 000 project partly funded by the EC. The idea is to incorporate some 2000 panels, each one foot by three feet, into the roof of a building so that they will supply up to 20 kilowatts of power in sunny weather – enough to provide power for lighting and equipment and computers within the building.

During the summer months there should be an excess of electricity which, they say, could be put to other uses within the building or sold to the grid. The cost of the photovoltaic panels should be virtually the same as ordinary glass cladding used in buildings and the company believes the technique has a great future.

In Phoenix, Arizona, there are already 24 homes whose electricity is supplied by photovoltaic panels, making it America's first solar electric suburban community. The cost is high and the electricity supplied is far more expensive than that from conventional sources. But within the next few years it is expected that the cost of photovoltaic panels could come down, making electricity from them comparable with nuclear electricity.

Despite the increasing interest in photovoltaics, the American Government, strangely, has actually been reducing its support of such research. Since the beginning of the eighties the US Department of Energy budget for photovoltaics has dropped from about $150 million a year to $35 million. Now the Japanese, German and Italian Governments spend more on this research than America. In 1980 US manufacturers had 80 per cent of the world market in photovoltaics, now it is less than 50 per cent.

The main thrust of Britain's programme on solar energy involves nothing so complicated as photovoltaics. It seeks to capitalise on the solar energy falling on buildings (passive solar energy) as a way of displacing fuel for heating, lighting and cooling. This is possible with little or no mechanical assistance – by putting the windows and skylights in positions where they let in most sunlight, for example. If solar power could be effectively tapped in sufficient quantities it would be the ultimate in renewable energy sources. That day, if it ever comes, is a long way into the future.

HYDROELECTRIC POWER

In 1986 hydroelectric power supplied 21 per cent of the world's electricity. That is less than the contribution from coal but more than from nuclear power. Among the biggest hydroelectric schemes is the Guri dam in Venezuela which can generate 10 000 megawatts – as much electricity as ten large nuclear power plants. Brazil is building a dam which should generate 20 per cent more electricity than the Guri dam and China is planning an even bigger one. But hydroelectric schemes do not have to be so large and there are thousands of smaller projects on remote rivers and streams which are supplying electricity, especially in developing countries. Here, particularly, there is scope for much greater development. While North America has developed 59 per cent of its hydroelectric potential and Europe has developed 36 per cent, Asia has harnessed only 9 per cent, Latin America 8 per cent and Africa 5 per cent.

Developing hydroelectric power is not without its disadvantages. Building dams floods vast areas destroying forests, farmlands and the habitats of wildlife, and displaces millions of people. Brazil has one of the largest programmes in the world having nearly tripled its generating capacity in the decade up to 1983. But many people think that too many rivers are now being dammed, with unfortunate environmental consequences. One scheme known as the Balbina will flood an area of 1554 square kilometres, altering the character of the land and climate and threatening many species with extinction. India plans to build 3000 dams in the Narmada valley project which would force a million people from their homes.

And once dams are built there can be unforeseen and expensive consequences. Reservoirs can silt up, reducing the efficiency with which they can generate electricity and shortening their life. The build-up of silt behind the Ambuklao dam in the Philippines is thought to have reduced its useful life by half. Huge reservoirs can become breeding grounds for the carriers of tropical diseases like malaria and river blindness. And downstream the rivers are robbed of rich silt which makes the flood plains so agriculturally valuable and feeds the fish. Damming rivers can severely interfere with activities downstream and that can lead to international tensions when rivers flow through several countries. Turkey's Ataturk dam on the Euphrates, for example, has led to lower water levels downstream and this has reduced the water supply to another hydroelectric scheme at Syria's Tabaq dam. Completed in 1978, it has five of eight 100 megawatt turbines now out of the water.

The World Bank estimates that nearly a quarter of a million megawatts of new hydroelectric capacity will have been added in developing countries between 1981 and 1995, more than half of it in India, China and Brazil. This figure is equivalent to 225 nuclear plants.

It is clear that hydroelectric power, properly managed, still has a huge potential – the key words here, however, are 'properly managed'. The Worldwatch Institute summed up the situation in a report: 'Hydroelectric power will not be truly renewable until the functions of flood control, irrigation, transportation, power production, tree planting, fisheries management and sanitation are coordinated with the overall goal of maintaining healthy and productive rivers.'

TIDAL POWER

Making use of the natural rise and fall of the tides to generate electricity has contributed a small but significant amount of electricity in certain areas. In theory it is an attractive idea, especially when the difference between high and low tide is considerable. As the tide floods in it fills a reservoir. The sluice gates are then closed trapping a large body of water as the tide goes out. This water is then allowed to flow out through turbines generating electricity. A three-metre tide can generate three times as much power as a one-metre tide. To be of economic use for electricity generation the tidal range needs to be between 3 and 5 metres. The world's greatest resource is at the upper end of the Bay of Fundy in Canada where the tidal range reaches 10–12 metres. Britain is considering a tidal barrage across the Severn estuary between Brean Down, west of Weston-super-Mare, and Lavernock Point near Cardiff where the tidal range reaches 35 feet. It would be a huge civil engineering project containing 216 turbines. It would cost more than £8 billion at 1988 prices but would produce between 6 and 7 per cent of Britain's electricity needs. If it were to replace coal- or oil-based electricity there would be a 17.6 million-tonne per year reduction in carbon dioxide emissions.

In Britain the Department of Energy estimates that tidal barrages could in theory produce the equivalent of 26 million tonnes of coal a year. Altogether some 34 potential sites have been identified in Britain and between them they could produce at least 100 megawatts of power from tides. They include the Thames, the Humber, the Mersey, the Wash, the Dee, Morecambe Bay and the Solway Firth. Evidence from France suggests that tidal barrages could pay for themselves in as little as 16 years. But such schemes would not be without their drawbacks, which include major environmental impact.

WAVE POWER

With the sea surface in constant motion it would seem a prime site for the deployment of devices which would capture the restless energy of the waves and convert it into electricity. In the middle of the North Atlantic, for example, the average power per metre of wave is about 90 kW. Nearer the coast it is less – about 25 to 70 kW. Clearly a huge amount of energy would theoretically be available to Britain particularly off the coasts of Cornwall and Scotland if the right devices could be developed. But there are major drawbacks, the key problem being to develop efficient and cost-effective wave power absorbers which can withstand storm conditions (during which the power per metre of wave can go up to 5 MW). A lot of devices have been developed and tested as scale models. Between 1974 and 1983 the Department of Energy in Britain funded a series of tests costing more than £17 million, which constituted probably the most exhaustive study of wave power devices in the world. In the mid-eighties, however, the Department concluded that wave power was not sufficiently attractive to warrant a large sum being committed to the next phase which would have been a full-scale test at sea. Recently though it has transpired that certain assumptions in the official assessment were wrong and wave power would probably be cheaper than officially thought. It seems set to make a new impact.

Internationally there are some projects which are being pursued with enthusiasm, notably in Japan and Norway, but projected costs of electricity produced in this way are high. Most interest is centred on shore-based systems which make use of what is called an oscillating water column – waves cause water in a chamber to rise and fall which in turn sends air through a turbine.

BIOMASS

Biological materials, either waste material like refuse in landfill sites or straw, or specially grown crops can be used as a source of energy. In the developing world biomass is the source of the largest proportion of energy used – some 43 per cent. World wide biomass contributes 14 per cent of total energy use. But proponents of biomass believe it has the potential to contribute far more.

Biofuels can either be burnt directly or converted into more conventional fuels like gas or alcohol. Small power units running on methane produced by the decay of urban waste in landfills are already providing a small amount of power. The United States is by far the largest user of landfill gas. The energy derived from it is equivalent to the energy from nearly 1.6 million tonnes of coal. Britain is second only to the US with West Germany coming third. In Britain the use of landfill gas saves an estimated 162 000 tonnes of coal a year and in West Germany landfill gas use is equivalent to 135 000 tonnes of coal. At the moment nearly 20 countries exploit landfill gas in one way or another either by using it for power generation as in America or by using it to fire kilns or boilers as in Britain (see Chapter 9).

Crops can also be grown specially for fuel. Brazil is one country where this is done on a large scale. Sugar cane is grown for alcohol production, for use in cars (see Chapter 7). Also, there is a lot of research being done into the use of wood as a biofuel. This research includes the evaluation and development of low cost techniques for producing, harvesting and using forestry products. Producing crops specifically for use as fuel provides an alternative use for agricultural land which, in the developed world, may not be needed for food production. And in the developing world growing crops for use as fuel could provide a stimulus for maintaining forests. Apart from providing fuel, such schemes also provide a sink for carbon dioxide.

GEOTHERMAL POWER

The heat in geological formations can provide an important source of power. In some parts of the world the hot rocks just beneath the surface of the earth naturally transfer their heat to water which flows to the surface, producing hot springs. In Britain the hot baths of Bath and Buxton originated in this way. In other countries like New Zealand and Japan there are more spectacular hot geysers.

Geothermal energy from naturally occurring hot water is exploited in some parts of the world, most notably in California and Iceland. In Reykjavik by 1975 all but 1 per cent of the buildings were connected to a district heating scheme using natural hot water. In France and Hungary, too, hot aquifers are used for district heating schemes but the temperature of the water rarely exceeds 100 degrees centigrade. The Italians were the first to use naturally occurring hot water to make electricity at the turn of the century. Now Italy produces about 400 megawatts in this way. Hot aquifers are also exploited in America, the Philippines, Japan and New Zealand. In fact, world wide it is estimated that some 4000 megawatts of electricity are produced from natural hot water.

The heat from hot rocks can also be tapped in another way where there is no water present naturally. Water or other more viscous fluid can be pumped down a specially drilled well under pressure so that it cracks the rock formation; alternatively explosives can be used. The energy from the hot rocks is extracted by pumping water down one well. The water heats up as it passes through the fissures in the rocks and then returns to the surface through a second well. To be of practical use the costs of drilling have to be kept to a minimum and there cannot be too much water loss through leakage. It was the 1973 oil crisis which stimulated research in Britain into energy from hot rocks and now British scientists are leading the research in this field. Britain has achieved longer and higher rates of water circulation through a hot dry rock reservoir than any other country in the world. The main test site is in West Cornwall where scientists have been working since 1977. Here, at a depth of just over a mile, the temperature of the reservoir is 75°C. It is really a research project to test the engineering involved as to be of any use for generating electricity, water has to be above 150°C. To reach that temperature the hot rocks have to be nearer 200°C and that temperature can normally only be found at depths of some 2½–3 miles. The Department of Energy in Britain believes that Cornwall's hot rocks may provide enough electricity in the next century to supply the South West – equivalent to some 10 per cent of Britain's electricity – for the next 125 years. The Energy Technology Support Unit at Harwell has in the past been more optimistic. It has said that technically Britain's resource of hot dry rock energy could provide 10 per cent of present electricity production for between 300 and 3000 years.

Whether geothermal energy will ever make a major impact

on energy supplies depends on its cost compared with other methods of generating electricity. Despite early optimism the outlook for hot dry rock energy production has begun to seem less promising than it was a few years ago. The practical difficulties and costs of operating such a system are proving less encouraging.

WIND

Of all the renewable sources of electricity wind energy is probably the most exciting. Compared with fossil fuels and nuclear energy wind power is relatively cheap, simple and clean. Aerogenerators, as wind energy machines are called, do not produce carbon dioxide or acid rain or gases like nitrogen dioxide which can be harmful to lungs. They don't produce radioactive waste. They can be built quickly and they can be dismantled quickly leaving no scars on the countryside. In areas where wind erosion is a problem, aerogenerators can in a sense 'capture' the wind and protect the soil from being blown away. They can be adapted in size to suit their purpose: a large aerogenerator can produce megawatts of electricity; a small machine could produce just enough electricity for a single household. In China, where large areas of the country are without electricity, a wind generator can be used to power a single television set linking those who live remotely with the rest of the world.

But the energy in the wind is weak compared with the energy packed into fossil fuels. To generate large quantities of electricity you need large areas of land covered with aerogenerators. Wind machines interfere with each other if they are too close together and they can be unsightly, though some might argue that we put up with miles of unsightly electricity pylons and at least wind machines can be concentrated in one area. They also produce a whirring noise and in some areas they can interfere with television reception. But these are minor problems compared with the problems of other forms of electricity generation. So how realistic is wind power?

Some indication of the terrific boom in interest in wind

power can be gleaned from the figures for sales of wind generators after the oil crisis of the early seventies. During the following decade more than 10 000 wind machines were installed world wide.

One place where wind power has made a big impact is California. Tax incentives led to a 'windrush' there in the early 1980s but it had unfortunate consequences. American turbine manufacturers did not have to obtain certification for their products so many jumped in with inadequately tested machines with the result that there was a high incidence of breakdowns and many firms went bankrupt. The reputation of wind energy was tarnished but many companies learned their lesson and now the market is once more healthy. The boom began in 1981 with the building of just 144 fairly small turbines generating a combined total of 7 megawatts of electricity. By the end of 1987 California had 16 661 turbines capable of producing 1437 megawatts of electricity. The electricity generated in California is sufficient to provide San Francisco with 15 per cent of its electricity demand. Enthusiasm for wind power waned when oil prices eased in 1986 but generating capacity continued to increase, if more slowly.

Most of the medium-sized wind machines in the world are in just three places in California – Altamont, San Gorgonio and Tehachapi. Although the idea of aerogenerators in this landscape with its sheep and cattle ranchers sounds incongruous, in fact the influx of wind power has provided an extra bonus for the landowners who lease out the land for the wind machines. Some twenty-nine square miles of the Altamont Pass are now covered with wind machines – one turbine for every three acres of land. Annual gross electricity sales per acre are worth about $6000, some 15 times more than the return a corn farmer in Iowa might expect, for example, and 100 times what a Texas rancher could hope for.

While it is estimated that wind power could in theory supply America with a quarter of its electricity demand at the turn of the century, Europe has an even greater potential because of its windy coasts.

Dominating the wind energy market is Denmark. The Danes

got off to an early start: wind generators were first developed there in the 1890s. Danish manufacturers produce reliable machines and supply a good proportion of the Californian market. In Denmark itself by the end of 1986 nearly 90 megawatts of wind power had been installed.

While most intermediate turbines produce typically about 55 kilowatts there are a few machines which can produce upwards of a megawatt of electricity. By 1985 seven had been built in the United States, two in Sweden and one each in Denmark and West Germany.

Probably one of the biggest markets is in the Third World. The Chinese Government foresees wind farms with a total capacity of at least 100 megawatts to be built between 1990 and 1995. In the Netherlands the goal is 150 megawatts by 1992.

Britain has been trying out a number of different types of wind turbines on Orkney and at Carmarthen Bay. Britain's interest in wind energy has reached the point where in 1989 the CEGB, as it then was, sought planning permission to build a wind park, or wind farm as it is called, in Capel Cynon in south-west Wales. It is designed to produce enough electricity to supply 5000 people and should be fully operational in the spring of 1991. To be of any use average wind speeds in any particular area must be at least 6.5 metres per second. Not all Britain has been surveyed to find the most suitable sites for wind machines but there are some obvious candidates: Cornwall, the Pembrokeshire coast, the Lleyn Peninsula, Anglesey, West Cumbria, Galloway, Kintyre and the West Coast of Scotland. The Hebrides, Orkney and Shetland are also good candidates. There are windy parts of the east coast and several suitable sites inland. Unfortunately many of the places are also sites of outstanding natural beauty. If you discount those and other areas where wind energy could not be exploited for various reasons, such as the difficulties of the terrain then the figures read like this: of the total land area in Britain of some 250 000 square kilometres about 5870 square kilometres might be useable – three-quarters of it in Scotland. That is sufficient to generate between 10 and 20 per cent of Britain's electricity supplies. The hills of Scotland are particularly windy – very useful for generating electricity, as a doubling of wind speed increases the amount of energy generated eight times. But Scotland is, of course, a long way from the centres where most of the power is needed and there are inevitable losses in transmission. To produce upwards of 10 per cent of Britain's electricity supplies would require between 10 000 and 20 000 wind turbines. That compares with 40 000 electricity pylons which already pepper the skyline.

Noise from aerogenerators is something of a problem, but normally it cannot be heard from more than 300 yards away. In the Netherlands proposed guidelines suggest that wind machines should not be sited within a distance of 300 yards from the nearest house. The European Community is to standardise guidelines by the end of 1990.

So how do the costs of wind power compare with other forms of energy? Recent figures show that wind machines can be built for £500 to £1000 per kilowatt. And they cost between 2 to 3 pence per kilowatt hour. Electricity from coal costs £1000 per kilowatt capital cost and from 2.8 to 3.4 pence per kilowatt hour. Nuclear power costs up to £1400 per kilowatt capital cost and (though these figures are subject to interpretation) 3–4 pence per kilowatt hour. Clearly then wind power must be considered one of the most environmentally friendly forms of generating electricity. In the end wind may never produce more than a small percentage of energy in Britain but it would be a vitally important percentage.

Exploiting renewable sources of energy is not necessarily cheap. Though the 'energy' is free the systems needed to tap that energy may well be expensive to build and maintain. Ultimately their contribution will depend not just on how easy they are to exploit but their costs relative to such sources as fossil fuels and nuclear energy. Even taken together renewable energy sources can satisfy only a relatively small proportion of the world's rising demand for energy. The proportion will increase but they will not by themselves solve the energy crisis which will eventually overtake the world once finite resources come close to being used up. One certain way in which this looming crisis can be slowed down is an improvement in the efficiency with which energy is used.

ENERGY EFFICIENCY

Buildings consume an enormous quantity of energy for heating, lighting and air conditioning. It has been calculated that in 1985 the buildings in industrialised countries consumed an amount of energy almost equivalent to the output of OPEC countries.

There is great scope for using energy more efficiently in buildings. Greater use could be made of so-called condensing furnaces which absorb a great deal of heat from boiler exhaust gases. These need 28 per cent less fuel than conventional gas boilers and produce fewer pollutants.

There are systems which can be incorporated into buildings which can sense where heating, lighting and air conditioning are needed. They monitor the environment outside the building, including measuring sunlight and so on, and can also tell where people are inside the building and deliver light, heat and air conditioning only to those areas where it is required. It is estimated these can produce energy savings of between 10 and 20 per cent.

Better building techniques and insulation can reduce heat lost through walls and windows. So-called 'super-insulated' homes where normal insulation is doubled and where there is an airtight seal in walls can save up to 68 per cent of the energy used for heating. In Sweden the savings can be nearly 90 per cent. In America research done at the Lawrence Berkeley Centre for Building Science has shown that for an investment of a mere $8 million it would be possible to produce and install so-called 'low-emissivity' windows for buildings which could save energy equivalent to $300 million.

Another energy-saving idea with a big future is 'cogeneration', or combined heat and power. Electricity generation in power stations is very inefficient – only a third of the energy from the steam is converted to electricity, the rest is lost. In a combined heat and power plant that energy is put to use for space heating and for producing hot water.

In industry, particularly in developing countries, there are many energy-saving techniques which can be introduced, thereby dramatically reducing energy bills. In China and India, for example, four times as much energy is used in making a ton of steel than in Japan. There is scope for improvement too in the production of electricity in ordinary power stations. One idea is based on a novel way of burning natural gas in a turbine similar to the way in which it is burnt in a jet engine. This is said to increase the efficiency of electricity production from the 33 per cent achieved in a conventional turbine to almost 45 per cent.

In Britain, Friends of the Earth and the Association for the Conservation of Energy have for a long time been stressing the importance of energy efficiency. Apart from saving energy they believe that if the technical and economic potential for energy efficiency were realised, pollution from existing methods of producing electricity could be cut by 70 per cent. They also believe that energy efficiency measures would be five to ten times cheaper than building new power stations.

Some countries are taking energy efficiency very seriously. Though American cars still consume far more petrol mile for mile than cars in most other countries, America does seem to have set the pace on ensuring efficiency in electrical appliances. Federal law now sets minimum efficiency standards for a range of new domestic products. These standards are so strict that up to 90 per cent of the electrical goods in the shops in 1986 would not be able to comply with the new standards when they come into force between 1990 and 1992. The new energy efficient appliances will save an estimated 21 000 MW of peak electricity demand by the year 2000 – equivalent to the output of about 20 large power stations. Other things being equal it will reduce the overall domestic electricity consumption by 6 per cent and by the year 2010 will be preventing the output of 70 million tonnes of carbon dioxide annually.

Friends of the Earth estimate that if similar high standards on electrical appliances were made mandatory in Britain about 90 per cent of the electrical goods here too would be effectively disallowed. To take an example: refrigerators and freezers account for nearly a fifth of all electricity used in the home. If they were all brought up to the standard of the best 10 per cent, savings of 600 megawatts a year could be made. Apart from saving electricity they would also be cheaper to run for the consumer.

If such savings are to be achieved there would have to be a commitment from the Government for minimum energy standards to be set, proper monitoring of developments, and adequate labelling so consumers know what they are buying. The Association of Manufacturers of Domestic Electrical Appliances, in a statement issued in response to that particular Friends of the Earth campaign, said they were against providing energy consumption figures on labels. The Association claimed energy efficiency was only one aspect of product performance and that consumers looked for other things as well when buying such products. Setting minimum standards, it claimed, might lead to what it called 'convoy technology' inhibiting further improvements in performance – a suggestion which left conservationists singularly unimpressed.

In another American scheme for improving the efficiency of electrical appliances, an electric utility subsidised customers to replace inefficient lighting with more efficient equipment and shared the savings.

But one concern in international terms is that the energy efficiencies being achieved in the industrialised world may be cancelled out by the increasing use of energy in the developing world as it seeks to match the West's standard of living. In fact there is no need for the Third World to be so profligate with energy as the West has been in order to achieve a high standard of living. A recent study which looked at this question concluded that by using the best technology available today it would be possible to provide a developing nation with a standard of living equivalent to that enjoyed by the developed world in the mid 1970s while increasing energy consumption in that country by only about 20 per cent.

A major problem is that market prices do not reflect true external and environmental costs, so government intervention is essential. This is one reason why industrialised nations help developing nations with subsidies, loans or technology transfer so that they need not rely solely on market forces.

CHAPTER FOUR

The Nuclear Option

Renewable energy supplies will certainly increase their contribution to the world's total energy consumption over the next few decades. The big question is by how much? Even with favourable research developments it is doubtful whether they will ever contribute more than a small proportion of the total amount of energy needed, one reason being that the energy sources are relatively diffuse, increasing collection and transmission costs. The only sort of energy generation which causes a minimal greenhouse effect and which concentrates sufficient power in one place to enable huge quantities of electricity to be made is nuclear power. The debate about the future of nuclear power has concentrated on two things – its safety and its costs.

HOW DOES NUCLEAR POWER WORK?

When a magnet is set spinning inside coils of wire an electric current is produced. In conventional power stations coal, oil or gas is burnt to heat up water in a boiler. The water turns into steam and the steam is then sent through the turbines to set them spinning. Once it has done its job the steam is condensed back to water in the giant cooling towers which are a feature of ordinary power stations. The water is then fed back into the boiler to be heated up again. A nuclear power station works in exactly the same way, except that instead of burning fossil fuels to heat the water the heat is produced by a nuclear chain reaction involving the substance uranium.

If the nucleus of an atom of uranium is hit by a sub atomic particle called a neutron travelling fairly slowly it captures the neutron. It now becomes less stable and may split into two, releasing a large amount of energy as heat and also releasing several more neutrons. These neutrons, particularly if they are slowed down sufficiently, can be captured by other uranium atoms so producing what is called a chain reaction. Material which can sustain such a chain reaction is called 'fissile'. Usually in a conventional reactor it is natural uranium (which is mostly the isotope uranium 238 with less than one per cent of the much more fissionable uranium 235), or so-called enriched uranium (in which the concentration of the isotope uranium 235 has been increased to about 2–3 per cent). In order to work best the neutrons have to be slowed down so they can be captured, and where this is done it is accomplished by the use of what is called a moderator.

In Britain's early reactors, the **magnox reactors**, graphite is used as the moderator and the heat is taken away from the reactor by carbon dioxide which passes through a series of pipes called a heat exchanger where it heats water to produce steam.

The **advanced gas cooled reactors** which make up Britain's second generation of nuclear power stations use enriched uranium and graphite as the moderator. They operate at a higher temperature and pressure so are more efficient. Again the heat is carried away from the reactor by carbon dioxide.

The world's most widely used type of reactor is the **pressurised water reactor** which uses more highly enriched uranium and is moderated by water which also acts as the coolant. It takes the heat away to a heat exchanger where it heats water in a separate circuit. There are several other types of nuclear reactors which work on slightly different principles. The 'next generation' of nuclear reactors may well be what are called **fast reactors**, in which the fuel is a mixture of uranium and plutonium (plutonium is made when a neutron is captured by a molecule of uranium 238). This fuel mixture, provided it is tightly packed, is able to sustain fission without the need for a moderator. The core of the reactor is smaller and therefore needs a very efficient coolant to take the heat away. Liquid sodium is used. Fast reactors can be surrounded by a blanket of uranium which captures stray neutrons and is turned into plutonium – hence the name of 'fast breeders' given to these reactors which reproduce their own plutonium.

Several countries are developing fast reactors. The French have a full-sized fast reactor called Super Phoenix and there is a European collaborative programme to build another one. But Britain has reduced its expenditure on fast breeder reactors and is winding down its prototype fast breeder reactor programme in Dounreay. To many this seems a retrograde step.

In all types of reactor the chain reaction can be stopped by inserting control rods into the core. They consist of a substance which absorbs neutrons and prevents any more fission occurring.

HOW SAFE IS NUCLEAR POWER?

This is a highly emotive subject nowadays, and there are several aspects of nuclear power which seem to be constantly under scrutiny: There is the debate about whether nuclear power stations can be operated safely without accident. There is the question of whether some aspects of nuclear power plants, such as 'normal emissions', can lead to an increased incidence of leukaemia among those who live near the plant. And finally, what should be done with the radioactive waste? Let us explore each of these points in more detail.

1 *Can nuclear power stations be operated safely?*

There have been three major accidents to nuclear reactors since the early 1950s, each of which has caused the industry to improve its safety precautions.

In 1957 there was a fire in the Number One pile at **Windscale** in Cumbria which was being used to produce plutonium for nuclear weapons. The reactors where the plutonium was being made were cooled by huge blowers which simply blew air through them and out through tall chimneys. John Cockcroft, one of the senior scientists at the Atomic Energy Authority, realised that if anything went wrong in the reactors and the uranium caught fire, highly radioactive material would be blown straight up the chimneys and into the air.

To avoid the possibility he fitted massive filters to the top of the chimneys. When the uranium and graphite in one of the reactors did indeed catch fire in 1957 the filters trapped most of the radioactive particles. Some radioactivity did escape into the environment leading to milk produced on farms nearby having to be thrown away. However, the fire was put out and the reactor sealed in concrete.

The first major accident to a commercial nuclear power station happened in America at the **Three Mile Island** plant in Middletown, Pennsylvania, close by the State Capitol of Harrisburg. It started at 4.00 am on 28 March 1979 when a fault caused a loss of feedwater to both the boilers. Three emergency pumps started up automatically, which should have fed water to the boilers, but all three valves had been shut off contrary to operating instructions. Because heat was not being conducted away from the core the reactor temperature and pressure rose. The reactor shut down automatically. A valve opened to relieve pressure building up in the primary circuit but did not close properly when pressure returned to normal and led to a continued loss of coolant.

It was at this stage that the second main cause of the accident occurred: the operators believed that the primary circuit was full of water even though it was losing water rapidly. That led them to shut off the emergency core cooling pumps which in turn resulted in the core heating up still further. This heat caused a reaction between the zirconium in the fuel rods and water, producing hydrogen which formed a large bubble in the water around the reactor core and prevented the top of the core being covered by water.

The various warning lights and gauges in the control room had been less than helpful to the operators. Eight minutes into the accident the lights on the control panel lit up 'like a Christmas tree', according to one of the operators, but those in the control room had no idea what it all meant. For two hours there was total confusion. Not one operator knew what the problem was nor what to do about it. It is believed that the temperature of the fuel rose to 1800°C and much of the fuel was damaged. The extent of that damage was not realised for months. Another problem was the length of time it took before anyone declared a site emergency, having realised that things were out of control. It was not until just after 7.00 am, when the situation was clearly very grave, that a supervisor called the Pennsylvania Emergency Management Agency and notified them of an official site emergency. But even then the situation was far from under control. The workers were evacuated in minibuses which proved inadequate for the job. Misinformation by television and radio reporters led the public to believe there might be a nuclear explosion and some 70 000–80 000 people voluntarily began to leave their homes and evacuate the area. Reporters and emergency units were trying to get in and residents were trying to get out. There was panic and confusion with no one in command of the situation. The veteran news reporter Walter Cronkite was later accused of adding to the panic by describing it as 'the first step in the nuclear nightmare'.

Despite the fact that it was a very serious accident very little harmful radioactivity escaped to the environment. According to the President's Commission on the accident, radiation doses received by the general population were so small that 'there will be no detectable additional cases of cancer, mental abnormalities or genetic ill-health as a consequence of the accident.'

But if Three Mile Island had frightened the nuclear industry internationally there was worse to come. At 1.23 am on 26 April 1986 an accident in the Number Four reactor at the Soviet Union's **Chernobyl** nuclear power station occurred which very nearly scuppered the nuclear industry world wide. It demonstrated for the first time how dangerous nuclear power stations can be if improperly designed and operated, it vindicated the warnings which the anti-nuclear lobby had long been voicing and it brought the nuclear industry itself face to face with a problem it had paid insufficient attention to. The accident occurred because the operators were carrying out an unauthorised experiment. They were trying to see if a free-wheeling turbine which was no longer being driven by steam could still generate enough electricity to keep the motors driving the reactor's cooling water pumps going for a short time. As the turbines slowed down so did the pumps until insufficient water was being sent through the reactor to keep it cool. The reactor overheated but the automatic safety system, which should have been triggered, failed to work because the operators had switched it off, although forbidden to do so. The increased heat caused the pressure of steam to build up until it burst the pressure tubes and lifted the pile cap leading to a massive escape of radioactivity.

It is calculated that 50 megacuries of radioactivity were released – equivalent to about 3.5 per cent of the total amount of radioactivity in the core. In the immediate aftermath of the accident two people were killed when they were buried in the rubble of the reactor, 28 people died of radiation sickness and more than 100 000 people were evacuated from a 30 km radius. The 24 000 people who lived between 3 and 15 km from the plant received doses of radiation between 350 and 550 mSv (compared with natural background radiation doses of 2.4 mSv). The rest of the evacuees received 30–60 mSv. Heroic efforts were made to save those people who had received particularly high doses of radiation. Bone marrow transplants were given to a number of patients but it is said this treatment did not save anyone and in some cases may well have hastened death through what is called the host versus graft reaction.

By October 1989, 250 people had died in the evacuated population. In fact that is the figure one would expect if they had not been irradiated, and their deaths cannot be definitely attributed to Chernobyl. However, an unknown number, possibly running into thousands, can be expected to die prematurely over the next 50 years from cancer caused by the extra radiation they received.

The plume of radioactivity spread over the surrounding countryside on the prevailing wind and in dilute form reached almost round the world. The legacy from Chernobyl has been chilling. For example, people in the Narodichi district 50 kilometres southwest of Chernobyl were not evacuated from the area after the accident even though the wind was blowing directly from Chernobyl. To give just one example of the consequences: in the year following Chernobyl at one farm with

350 cows and 87 pigs, 64 seriously deformed animals were born. During the first nine months of 1988 there were a further 41 deformed pigs and 35 deformed calves born. Some of the calves had no heads or limbs or eyes and most of the deformed pigs had abnormal skulls. Radiation levels at the farm were found to be 148 times higher than normal background levels. Though food is brought in from outside, people still drink milk from their own farms and grow their own vegetables, and the cattle graze on contaminated grass.

At the medical centre at Kiev, women from the area are being advised not to have children and there seems to be an increase in chronic illness. People are said to be recovering from illnesses and minor surgery more slowly than they ought. According to one report cancer cases have doubled.

Those who support nuclear power while not denying the disastrous consequences of Chernobyl point out that other forms of energy generation also cause death. For example, in January 1990, the US Government stated that exhaust fumes cause 100 000 emphysema and lung cancer deaths a year in the USA – a consequence of using oil as the energy source for transport. In 1988 an explosion in a coal mine in Germany killed 57 miners, a gas explosion in Mexico left 450 dead and thousands injured, in 1979 a dam burst in India killing 15 000 people. They point out that there will be many cases of cancer following Chernobyl over the next 50 years but that the natural occurrence of cancer is so high that the extra deaths will probably not be identifiable. None, though, excuse what happened at Chernobyl.

The Chernobyl-type reactor, known as an RBMK reactor, is an inherently unsafe design. As the water boils inside the reactor and is replaced by steam the reactivity of the nuclear core increases and power increases. This run-away tendency (known as a positive void coefficient) is meant to be controlled by rods which can be inserted from above and below to slow down the nuclear chain reaction. But in the reactor as it was originally designed, it took 18 seconds for the control rods to be fully inserted and they could not be inserted quickly enough to stop a power excursion. In addition another design fault was that there was no overall containment to prevent any fission products escaping in the event of an accident. The design is unique. Western nuclear experts say they had long been suspicious of this type of design. Britain decided in 1947 not to use it, fearing just such an accident. Yet, it has to be said, the West did little to persuade the Soviet Union of the potential dangers. The Chernobyl disaster, then, was caused by poor design, poor management and operating errors – a lethal combination.

But it was more than that – it was also the product of a strange management system. The director of the power station was apparently responsible not just for the reactor but also for everything to do with the site, including the workers' welfare. He found himself trying to organise repairs to the heating systems in their houses and even trying to obtain supplies of fresh vegetables for them to eat. He has revealed only recently that without breaking the rules Chernobyl would never have been built. He was worried about the situation, but he knew that if he made a fuss he would simply be replaced by a new and more amenable director. He tells of how he first learned of the accident. He was telephoned at home after the accident and told there had been an explosion. Thinking that some steam pipe had burst he hurried to the plant. He says when he saw that the top was missing from the Number Four reactor 'my heart stood still'.

Following the accident the Soviet Union implemented a range of safety measures to its remaining ten RBMK reactors. The control rods, even when withdrawn, now remain partially inserted into the core reducing the positive void coefficient and enabling them to be replaced more quickly. And an extra safety system has been installed which injects a radiation absorbing material into the core if it starts to over-react. Improvements in the training of operators have also been made.

Western nuclear agencies did their best to distance their own nuclear programmes from the Soviet one, emphasising that the type of accident which happened at Chernobyl could not happen in the West. Nevertheless the effect on the industry world wide was traumatic.

One good thing that came out of Chernobyl was the setting

up of what is called the World Association of Nuclear Operators. Every country in the world which has a nuclear power station belongs to WANO. The idea is that, should any unexpected event occur on any reactor, the owners of that reactor undertake to inform all other reactor owners of what has happened so that any lessons learned can be shared internationally. It is said that had WANO been in existence before Chernobyl then the accident would never have happened.

The accidents at Windscale, Three Mile Island and Chernobyl are not the only major nuclear accidents there have been, though they are the best known. There was a very serious accident at a high level nuclear waste facility in the Soviet Union in 1957, the existence of which was not officially admitted for many years and the details of which were not officially revealed in full until more than 30 years later. The accident took place at **Kyshtym** several hundred miles east of Moscow. An estimated 20 million curies of radioactivity were released (compared with 50 million released following Chernobyl) and 15 000 square kilometres were contaminated. It has been calculated that those initially evacuated from the area received between 23 and 500 mSv of radiation (the internationally agreed limit for exposure to radiation by a member of the public is 5 mSv a year). The explosion occurred in a concrete storage tank containing 160 tonnes of high level radioactive waste from the production of nuclear warheads. Clearly scientists and engineers know a great deal more now than they did in 1957 about the risks of nuclear accidents, and consequently guard against them more effectively, but any assessment of the overall safety of nuclear power cannot ignore history.

2 *Can the discharges from nuclear power stations and reprocessing plants cause leukaemia and other illnesses?*

This is one of the fundamental questions over which argument has raged for many years. The anti-nuclear lobby have long maintained that discharges from nuclear plants are probably causing an increased number of leukaemia cases, but official reports have not, until recently, been able to link any increase with radiation.

Studies of the possible risks of living near nuclear plants

have been pioneered by Britain. In 1983 a committee under Sir Douglas Black was commissioned by the Minister of Health to investigate reports of a high incidence of leukaemia in young people living in the village of Seascale in West Cumbria. Seascale is 3 kilometres from the Sellafield reprocessing plant belonging to British Nuclear Fuels and there was a suspicion that discharges from Sellafield might be responsible for the increase. The Black report confirmed that there was a higher incidence of leukaemia in young people living in the area compared with the average for England and Wales. But it also said that the estimated radiation dose received by the local population as a result of the discharges from Sellafield could not account for the increase based on current knowledge.

There were still some uncertainties and so to investigate the matter further the Black report recommended the setting up of the Committee on Medical Aspects of Radiation in the Environment. In one of its early reports the Committee extended its investigation to look at the incidence of leukaemia in young people near the Dounreay nuclear plant in Caithness in Scotland. They found evidence of an increased incidence of leukaemia in young people living in that area too although, once again, they found that neither authorised nor accidental discharges could have accounted for the difference.

In its third report the Committee looked at the incidence of childhood cancer in an area which included the Atomic Weapons Research Establishment at Aldermaston and the Royal Ordnance factory at Burghfield where nuclear weapons are made. It concluded that there had been a small but statistically significant increase in childhood leukaemia in the 0–14-year-olds between 1972 and 1985 in the areas within 10 kilometres of the nuclear sites, compared with the national rates. It found no increase in the rates of childhood leukaemia around the Atomic Energy Research Establishment at Harwell. The Committee also looked at the atmospheric and liquid discharges of radioactive material from the plants and found they were much too low to account for the increase in childhood cancer in the area, but they added, 'We cannot exclude completely the existence of some hitherto unknown and unexpected route by which some individuals could be exposed to increased levels of radiation. Such speculative pathways, including those involving radiation workers, should be explored.'

There have been claims too that nuclear power stations themselves may put young people living nearby at increased risk of leukaemia. But there is no evidence of any increased incidence of the disease. The average doses of radiation received by members of the public living near a nuclear power station would be only 2 per cent above normal background radiation. That represents less than the natural change in background radiation that a person would receive in moving from one part of the country to another and is equivalent to one return flight to the Mediterranean.

In the long-running debate about whether living near certain types of nuclear plants constitutes a risk, various causes apart from radiation have been put forward as possible reasons for the increased incidence of leukaemia. They include chemical carcinogens (cancer-causing chemicals), demographic phenomena and viruses. The Committee say in one of their reports: 'Although we recognise the considerable importance of these factors we are not aware of any specific evidence that these are responsible for the increased incidence of childhood cancer.' So despite several careful analyses of the incidence of leukaemia near certain nuclear sites there is still no explanation of why it should be raised.

Clusters of leukaemia cases also occur in areas of the country which are far removed from nuclear sites and certainly they could not have been caused by man-made radiation. One theory suggests that in a remote area it is possible that the level of immunity is slightly lower than in more crowded areas where germs have been exchanged more frequently and at a younger age. If leukaemia is caused by a virus then it is possible that young people living near remote nuclear sites might be more susceptible to infection following an influx of people. This kind of thing is commonplace but the evidence that it accounts for leukaemia is slim.

A report published in November 1989 revealed that the incidence of leukaemia and another form of cancer known as Hodgkin's disease in young people living in areas which had once been considered as possible nuclear sites but where no nuclear plant was built was similar to the incidence in areas where nuclear plants had actually been built. The scientists report: 'our hypothesis that, with the possible exception of Sellafield, an increased risk of leukaemia is not associated with environmental radiation pollution is strengthened by our new findings.' They add, 'our new findings point to systematic differences between districts near existing or potential installations and other districts with respect to some important, unrecognised risk factors.'

In February 1990, a team led by Professor Martin Gardner which investigated the incidence of leukaemia round Sellafield found that men who were exposed to 100 mSv or more of radiation had a six- to eight-fold increased risk of fathering a child who developed leukaemia. The researchers also found evidence that the children of men exposed to lower doses of radiation within six months of their children's conception were at increased risk, though the numbers in the study were small. A similar explanation did not seem to hold for the cluster of leukaemia cases near Dounreay in Scotland though more research is being done here and at the nuclear weapons plants at Aldermaston and Burghfield.

People are exposed to radiation all the time from the sun, from the soil and even from the air they breathe. On average the biggest source of radiation comes from radon gases present in the air. Radon is given off by rocks and soil and even by some building materials and can accumulate in homes. Radon and its associated gases provide 37 per cent of our total radiation dose. Direct radiation from uranium and similar elements in the soil makes up 19 per cent of our exposure, radiation from radioactive elements in our own bodies like potassium and the food we eat and drink contributes 17 per cent and radiation from outer space makes up 14 per cent. The rest of our exposure comes from things like medical X-rays and fallout from nuclear weapons when they were tested in the atmosphere. In addition, the average individual receives 0.5 per cent of his total exposure to radiation from the burning of coal, and travel by air. The nuclear industry contributes just 0.1 per cent of the average dose to an individual. These apparently comforting figures are frequently attacked by anti-nuclear groups on the very reasonable grounds that they take no account of the fact that for members of the public living near nuclear sites the relative contribution of radiation from the sites is that much higher. Even so, with the possible exception of Sellafield, there is no evidence that these routine discharges have caused any cases of leukaemia.

WHAT IS RADIATION?

The risk from radiation depends on the type of radiation under discussion. Atoms are usually stable but some are unstable and can change into other forms. They are radioactive and give off packets of energy either as rays or particles. The amount of damage this energy can do depends on how heavy the particle is and how fast it is travelling.

Alpha particles are heavy but slow and can easily be stopped – they cannot penetrate skin, for example, which is why plutonium which gives off **alpha radiation** can be held in the hand. But if ingested, alpha emitters are very dangerous because they are very long lasting.

Beta radiation consists of particles which are fast but very light. They are more penetrating than alpha particles but can be stopped by a sheet of aluminium.

X-rays and gamma rays are not particles but what are called **electromagnetic radiations**. They are like light and radio waves but much more penetrating. They travel at the speed of light and can pass through aluminium but can be stopped by lead. Finally the most penetrating radiation of all is radiation due to neutron particles. **Neutron radiation** is only released inside nuclear reactors and though neutrons can pass through lead they are stopped by concrete.

Nuclear waste gives off all types of radiation, except neutrons, to various degrees depending on what the waste consists of. Another aspect of radiation, which is important when considering the question of nuclear waste, is known as the half life of a radioactive substance. The half life of a radioactive chemical is the length of time it takes for the chemical to lose half its radioactivity. Half lives vary from a fraction of a second to thousands of millions of years. In decaying, some radioactive chemicals can turn into other unstable types with their own half lives. Plutonium and some of the fission products have very long half lives. Others, like radioactive iodine, which is also produced in reactors, have half lives measured in a few weeks.

3 *Which leads us to our third problem: how to dispose of nuclear waste.*

Nuclear waste comes in three forms – low, intermediate and high level waste.

Low level waste includes things like protective clothing, glassware and swabs. Britain produces 25 000 cubic metres a year of low level waste. In ten years' time there will be some 500 000 cubic metres of low level waste in Britain awaiting disposal. By the year 2030 this will have increased to 1.5 million cubic metres.

Short lived intermediate level waste is 1000 times more radioactive than low level waste. It needs shielding and remote handling and it includes things like resins and air filters – objects which have not been in contact with nuclear fuel. Less than 1000 cubic metres are produced in Britain each year. By the year 2000 some 25 000 cubic metres will await disposal.

Long lived intermediate waste includes things like fuel cans from reprocessing. 1500 cubic metres are produced each year and by the year 2000 there will be 55 000 cubic metres awaiting disposal.

Highly active fission products are the most dangerous nuclear wastes of all. They are separated from the useful uranium and plutonium during reprocessing. If reprocessing is not done the used fuel rods themselves would be regarded as high level waste. They contain 99 per cent of all the radioactivity connected with nuclear power stations.

High level wastes in liquid form are building up at the rate of 100 cubic metres a year. They too are stored at Sellafield in specially cooled tanks where the liquid is kept constantly stirred.

In Britain the plan at the moment is to dispose of the low and intermediate level wastes in a repository at least 300 yards deep underground. Originally the low level waste was to be disposed of in shallow trenches but finalising a site for such a depository from a choice of four – all in what were Tory constituencies – in the run up to the 1987 election stirred up such a controversy that the Government decided to abandon those plans and put all low and intermediate level waste deep underground. The extra cost of burying so much relatively harmless low level waste so deep was apparently worth the political return.

In Britain the high level waste is to be stored for fifty years until it has cooled and then it will be turned into blocks of glass – so-called vitrified waste – for disposal again, probably down a deep hole. The decision to postpone doing anything about this high level waste for fifty years has neatly avoided the problem of high level waste becoming a political issue. Other countries are committed to finding a permanent answer to high level waste long before Britain. In Germany and the United States there are plans to begin disposing of high level waste in the early 2000s; in France by about 2010, and in Belgium, Canada, Finland, Japan, Sweden and Switzerland by about 2020.

The operation of all the nuclear plants in the world in 1988 gave rise to some 7000 tons of spent fuel. If all the electricity generated by that fuel had been generated by burning coal it would not only have resulted in the emission of 1600 million tons of carbon dioxide and tens of millions of tons of sulphur dioxide and oxides of nitrogen which contribute to acid rain but in addition 100 000 tons of poisonous heavy metals including arsenic, cadmium, chromium, copper, lead and vanadium. They are not isolated from the environment as nuclear wastes are and they do not decay.

There is another aspect of nuclear waste to be considered – the decommissioning of time-expired nuclear reactors. Experience world wide so far is limited. The largest reactor yet to be decommissioned is the tiny 22 megawatt Elk River reactor in Minnesota. It only operated for four years yet it took three years to dismantle in the early seventies. A bigger reactor, the 72 megawatt Shippingport plant near Pittsburg, could have been used as a test bed for decommissioning techniques but instead the steel reactor vessel was encased in concrete, transferred to a barge, sent down the Ohio and Mississippi rivers, through the gulf of Mexico and the Panama Canal and then up the Pacific coast and the Columbia river to be buried in a trench on the Government-owned nuclear reservation in Hanford, Washington.

The world's first full-sized nuclear power station to be decommissioned after completing its useful life will be the Berkeley power station in Gloucestershire. The decommissioning will take place in three stages. In stage one all the fuel in the reactor will be removed. This sounds simple enough, but it is a job which will, in fact, take five years. The spent fuel will be sent to Sellafield for reprocessing. Stage two will take up to a further seven years and will involve the removal of all the plant and buildings except the boilers, pressure vessels and the biological shield. The station could then be sealed up and left for a hundred years, during which time the residual radioactivity inside the reactors will have decreased by 100 000 times. At that time the reactor could be demolished by men working inside it. That would save having to use remote control equipment. Whether it is feasible to seal and guard a place for a hundred years and guarantee that it will not deteriorate to a dangerous extent or suffer some sort of interference is a matter for debate. The likelihood is that the site would still be needed for the generation of electricity by whoever owns it, so some form of official presence would be virtually guaranteed. It may not, therefore, be as unreasonable as it might seem to have these huge concrete nuclear tombs standing unused for a century. After all, the contaminated Number One pile at Windscale has remained sealed since the accident in 1957 and no one questions how reliable that policy is.

Once the radioacivity has decayed, disposing of the concrete blocks would not be particularly difficult. An original option was to drop them into the deep ocean. But in 1983 the London Dumping Convention which controls dumping at sea decreed a moratorium on this type of disposal. Instead it is likely the concrete blocks and the other contents of the reactor will be disposed of deep underground.

Another frequently discussed point is the cost of decommissioning. This depends to a large extent on how fast it is done. Leaving the radioactivity to decay for 100 years – the option of choice by Britain's nuclear industry – is not cheap because during that time the entombed reactor has to be permanently guarded. One estimate suggests that decommissioning a reactor may well cost £300 million. But estimating costs 100 years ahead must be a fairly pointless exercise. Other countries intend to decommission their reactors sooner than this – probably a more realistic option for two reasons. Firstly, it is politically more honest. Britain does seem to have tried to avoid the awkward question of both the disposal of high level waste and the complete decommissioning of reactors by the expedient of simply postponing the decision for a generation or two. Secondly, it is surely immoral to saddle generations yet unborn with having to clear up the mess we will have left behind. It is doubtful they will thank us any more than we thank those of past generations who have disfigured the countryside with the spoil from mines.

Choosing a site and organising the safe disposal of waste, does not, according to the nuclear industry, represent any special technological challenge. Engineers know how they can achieve the goal of separating nuclear waste from man's environment. They are also confident the waste will not return to man's environment for tens of thousands of years. But what has held up the programme in Britain and elsewhere is the public's unease about the policy. Nuclear power has been used by mankind for only about 33 years. Yet already a fairly sizeable amount of waste has been generated and stored and some waste has been discharged to the environment. If there are to be no new developments in energy generation over the next century or so, and mankind has to continue to rely on reactors based on nuclear fission, then the amount of nuclear waste will continue to increase. It is entirely reasonable to assume that all the waste could be disposed of in deep geological formations, but inevitably the more there is the greater is the chance that some of the waste will contaminate man's environment.

Such was the enthusiasm with which several governments embraced nuclear power, initially as a way of producing material for weapons, that little thought was given to the problem of what to do with the waste. In Britain, doubts were first publicly addressed in 1976 by the Royal Commission on Environmental Pollution which concluded one of its reports by saying 'it would be irresponsible and morally wrong to commit future generations to the consequences of fission power on a massive scale unless it has been demonstrated beyond reasonable doubt that at least one method exists for the safe isolation of these wastes for the indefinite future.' In the eyes of many, the industry has indeed addressed these issues and found that there is no reason to suggest nuclear wastes cannot be handled safely.

Certain environmental lobby groups have campaigned for nuclear waste to be left where it arises as they are concerned about the dangers inherent in transporting waste to the single disposal site. It is hard to see any justification for this policy. It would mean leaving nuclear waste in store, probably on the surface, at several hundred sites in Britain, possibly more. The costs of guarding the sites would be large, and the risks for those workers at the site would be increased as would risks to the public living nearby, should an accident happen which might disperse the wastes. It hardly sounds sensible, even given that it would be only a 'temporary' measure until a permanent solution could be found. It is most likely that, had the nuclear regulatory authorities themselves suggested such a scheme, environmental organisations would have been the first to say it was a dangerous and short-sighted policy. In these circumstances one can easily imagine them campaigning for a single deep disposal site.

HOW MUCH DOES NUCLEAR POWER COST?

The issue of the cost of nuclear generated electricity is one which has caused endless arguments over the years between the industry, which claims that nuclear electricity is cheaper than that produced by conventional power stations, and the anti-nuclear lobby which claims it is more expensive.

Figures produced by the International Union of Producers and Distributors of Nuclear Energy in 1988 showed that in seven European countries and Japan nuclear power will be 38 per cent cheaper than coal-fired electricity by 1995.

But calculating nuclear electricity costs is especially difficult and the startling events of 1989 in Britain, when the Government withdrew nuclear power stations from its privatisation plans, served only to show that reassurances on the cost of nuclear electricity can carry little weight. The cost comparisons with coal showed that, after all, nuclear power is not such a good buy.

Clearly everything depends on what is included in the costs. The figures given at the inquiry into the building of the Pressurised Water Reactor at Hinkley Point in Somerset are a good starting point. The CEGB said that at 1987 prices nuclear power would cost 2.2 pence per kWh compared with a cost of 2.6 pence per kWh from a coastally sited coal-fired power station which could use cheap imported coal. In 1989 prices that would be 2.5 pence per kWh for nuclear and 3.00 pence per kWh for coal. These figures were based on nuclear power stations built in the public sector paying interest at 5 per cent real rate of return. Were the nuclear power stations to be privatised the banks would want a bigger rate of return – say 10 per cent. That would put up the cost of nuclear electricity to 3.8 pence per kWh compared with 3.4 pence for coal. Since those figures were given there has been a 10 per cent increase in the cost of building the Sizewell Pressurised Water Reactor. If that increase is reflected in the cost of nuclear electricity it would put up the cost to 4.2 pence per kWh compared with an unchanged 3.4 pence for coal. In adddition, the costs could well go up still further if the banks decided they wanted their capital returned over 20 years rather than 40 years. If increased estimates of the costs of decommissioning are included in the equation alongside the increased costs of reprocessing, nuclear power comes out a very poor second. And finally there is the uncertainty factor: a private nuclear industry might well be bankrupted if there was a Chernobyl-type disaster, whereas a government-owned nuclear industry would not be.

From these figures it is clear that the prices charged by a privatised nuclear industry would have to be very much higher than prices charged by a state owned industry. The actual figure was given at the annual lecture of the British Nuclear Energy Society by Lord Marshall of Goring, Chairman of the Central Electricity Generating Board, who resigned once nuclear power stations were taken out of the privatised electricity industry. He said that the final price would have to be 6.25 pence per kW – considerably more than the price of other forms of generation.

In France the costs of nuclear power are different again. The French have a large number of nuclear reactors of a similar type so they have benefited from the savings to be had from 'mass production'. France also has no domestic coal or oil resources to speak of and so had little option than to go nuclear in a big way.

But the costs of electricity, too, depend on what you include in the calculations. If nuclear costs have to include the costs of decommissioning and waste disposal then one should also include the *indirect* costs of fossil fuel power stations. Top of the list here should, of course, be a proportion of the cost of the greenhouse effect. The costs of combating even a moderate rise in sea level and of changing agricultural patterns could well be astronomical and that would put the cost of electricity from fossil fuels way beyond that of nuclear power. After all, most scientists accept that some form of greenhouse effect is a certainty if we continue burning fossil fuels, whereas the possibility of a major nuclear accident like Chernobyl remains just that – only a possibility – even though its financial consequences would be huge.

Although the nuclear construction industry world wide has suffered a serious slump since Chernobyl, by the end of 1988 there were no fewer than 428 nuclear power reactors in operation around the world producing an estimated 309 564 megawatts of electricity. Between them they have clocked up over 5000 years of operating experience. There were a further 106 reactors being built which are expected to produce another 87 569 megawatts when completed. The United States has the most nuclear reactors – 108, while the Soviet Union has 57, France 55 and Britain 40 (most nuclear power stations have two reactors each). In Germany nuclear power contributes nearly 40 per cent to its total electricity production, in Belgium, 65.5 per cent and in Spain, 36 per cent (and there is speculation that the Spanish Government may lift its ban on new nuclear plant construction). In France nuclear power makes the biggest contribution of all – no less than 70 per cent. France also produces electricity for export, especially to the UK. Nuclear power now accounts for over 16 per cent of the world's electricity production. If the world's nuclear electricity had to be replaced by electricity from fossil fuels it would mean the burning of more than 900 million tons of coal or 600 million tons of oil a year, resulting in the emission of many millions of tons of carbon dioxide into the atmosphere. If the number of nuclear power stations were to double over the next fifteen years carbon dioxide emissions might be cut by 15 per cent – three-quarters of the total cut called for by the 1988 Toronto conference on the changing atmosphere.

It has often been said that to replace coal-fired power stations with nuclear plants would require one new nuclear plant to be put into operation every 2.5 days for the next 35 years. That sounds a ludicrous goal but the argument falls on two counts. No one is suggesting that all coal-fired plants will have to be replaced – a mix of energy supplies is essential for many reasons. And it is a fact that in both 1984 and 1985, 33 new nuclear plants came into operation around the world – that amounts to one every 11 days, and the capacity exists to build many more nuclear plants than are built today.

There is no doubt that energy use is set to grow substantially over the next 30 years. Even based on figures showing a fairly low rate of growth, it is estimated that electricity demand around the world will double by the year 2020. That would require an additional 1000 MW plant being brought into operation every 4.4 days on average, and there would be further demand for the replacement of old and obsolete plants. If this extra energy were to be provided by coal-fired plants a further 16 500 million tons of carbon dioxide would be emitted into the atmosphere each year.

One of the consequences of the Three Mile Island accident in the United States was that no new nuclear power stations have been ordered since then. As a result energy analysts there are forecasting that America could find itself seriously short of electricity in the next few years especially in the eastern half of the country. In October 1989 the United States Energy Secretary James Watkins warned that declining electricity generating reserve margins jeopardise the economic future of the United States. In addition, there have also been warnings that unless the United States reconciles its energy and environmental concerns, energy security will deteriorate to the point where the nation would be importing up to 65 per cent of its oil, which would leave the United States particularly vulnerable.

One country which the anti-nuclear lobby often cite as an example to the rest of the world to show that public opinion can force a country to phase out nuclear power is Sweden. In 1980 there was a referendum on nuclear power which decided that nuclear power should be ended. Opponents of nuclear power proposed an immediate halt to the construction of new nuclear power stations, and the closure of operating plants by 1990 (later amended to 2020). They claimed that energy conservation, solar, wind and biomass power could be used instead. But by mid-1989 Sweden had used 35 per cent *more* electricity than it did in 1980 and about 45 per cent of it came from nuclear power. (Fifty per cent of it came from hydro power with solar and wind power providing just 0.004 per cent.) In January 1990 there was a cabinet reorganisation in Sweden which signalled a further retreat on its plans to phase out nuclear power. One of the keenest supporters for a phase out, the energy minister Mrs Birgitta Dahl, was replaced. The plan had been to shut down two of the country's 12 nuclear stations by 1996.

* * *

IS NUCLEAR POWER ON THE DECLINE TO AN IRREVERSIBLE LOW POINT?

Few would deny that the increased use of nuclear power throws up some very difficult questions. But most scientists and engineers would argue that there are no insurmountable technical problems either with making nuclear stations safer to run or in disposing of the waste, whether or not reprocessing is undertaken. Some would argue that the increased use of fossil fuels poses even greater problems. There is a chance, even if some would say it is remote, that the world could be faced with another accident on the scale of Chernobyl or worse. But it is just that – a chance. Compare that with the certainty that if the world continues to pump out carbon dioxide into the atmosphere at an ever increasing rate then the greenhouse effect will manifest itself sooner or later with international repercussions which could outweigh any threat posed by nuclear power.

Talking of the greenhouse effect at a conference on nuclear power in London in 1989, Dr Hans Blix, Director General of the International Atomic Energy Agency, said: 'We cannot plan substantial reductions in the use of coal, oil or gas on the basis of dreams. We need to escape from the greenhouse, yes, but we need also to escape from the dreamhouse.'

The arguments in favour of expanding nuclear power have yet to satisfy a large body of opinion, particularly in the United States. There is, however, growing interest in reactors of different designs developed from existing reactors which would in theory be inherently safer. They could satisfy existing safety criteria by relying more on the laws of nature to make the reactor safe if anything went wrong rather than on engineered safety features which themselves may go wrong. Most of the new designs embody two principles. First, they can be kept cool by natural convection and, second, if the reactor temperature rises the chain reaction dies down spontaneously.

There are a number of ideas being explored. One type of reactor uses different coolants – gases like helium which are said to be less corrosive than the carbon dioxide used in existing gas-cooled reactors. The new design of reactor uses fuel made in a different way so that it is dilute and can withstand high temperatures without failing. The fuel is embedded in blocks of graphite and this construction helps trap radioactive products. One disadvantage of this type of reactor is that it can only be built on a small scale and consequently could be uneconomic. But from another point of view building a power station in modular form using these reactors could well be attractive economically as it would allow for expansion as the demand for electricity increased.

There are also new designs for the fuel. Instead of the old fashioned ceramic fuels the new reactors would use alloys of uranium, plutonium and zirconium. The idea behind this is that if the coolant were lost the fuel would heat up and expand. This would allow neutrons to leak out, the chain reaction would grind to a halt and the whole system would cool spontaneously.

Another type, known as the Process Inherent Ultimate Safety type, or PIUS, is a Pressurised Water Reactor in which the core, the cooling system and the steam generators are all immersed in a solution of boron which is a neutron absorbing material. The idea is that when the reactor is operating normally the pressure from the pumps keeps the boron out of the reactor. In the event of the pumps failing the boronated water would come flooding in under the force of gravity and close the reaction down.

In Britain the Atomic Energy Authority is designing what it calls a Safe Integral Reactor, or SIR, and plans to build one at Winfrith in Dorset. It is basically a Pressurised Water Reactor but simpler than the type being built at Sizewell and about a quarter of the size. It would have an output of 300 megawatts. It is called 'Integral' because the core, the steam generators, pumps and pressuriser are all placed in one compact pressure vessel in a concrete-lined cavity. If anything were to go wrong the heat generated in the reactor would in theory be conducted away 'passively' by the natural convection of the water and would not have to rely on pumps to keep the cooling water flowing. It would have a different sort of fuel so that as the temperature increased the fuel itself would begin to absorb more neutrons. If the temperature became too high then the chain reaction would slow down and stop. The SIR could, it is believed, take only 30 months to construct – half the time of a Sizewell-type Pressurised Water Reactor, and would be based on existing technology so it would not involve a long lead time for research and development with all the problems that can cause.

These developments may well commend themselves to public opinion so that, should it eventually become apparent that demand for electricity cannot be satisfied with conventional power stations, the nuclear option can be presented afresh, together with reassurances which would make accidents like Three Mile Island and Chernobyl virtually unrepeatable.

The major drawback to all thermal reactors is that they use uranium very inefficiently so that only the richest uranium deposits can fuel them economically. World wide these deposits are equivalent to an energy resource of only half that of the world's oil. Some experts believe that by the year 2050 fuel for the world's PWRs may be prohibitively expensive. By that time thermal reactors would have to give way to fast breeder reactors which use uranium 60 times more efficiently than thermal reactors.

Fast reactors will almost certainly bridge the gap between thermal reactors and the ultimate type of nuclear reactor – the fusion reactor.

While thermal reactors depend on the release of energy produced when heavy atoms of uranium are split, fusion reactors depend on the energy released when two light atoms – isotopes of hydrogen called deuterium and tritium – fuse. This is the process which powers the sun. Should a fusion power station ever become a reality it would have a number of important advantages over conventional nuclear power stations. It would have an almost unlimited supply of fuel – hydrogen present in sea water. It does not produce the sort of dangerous fission products which thermal reactors produce so there is much less radioactive waste. And it is inherently safe. The conditions required to produce a fusion reactor are such that the slightest disruption would simply stop the reaction.

To achieve fusion it is necessary to heat the hydrogen isotopes to a temperature of 100 million degrees, at which stage they become a plasma (an electrically neutral mixture of ions and electrons), and to hold them at that temperature long enough and densely packed enough for fusion to take place. The European Fusion research programme based at Culham in Oxfordshire has reached all three targets, but not simultaneously. They now believe that it is only a matter of time before they produce the conditions required to make a fusion reactor – one that would generate more electricity than it consumes. Clearly it would be a big jump from achieving that to producing a commercial nuclear power station but the advantages are so great it would seem inevitable that one day it will be done. Already countries with an interest in fusion power, the European Community, the USA, Japan and the Soviet Union have agreed on the basic technical description of an experimental facility that could pave the way for a fusion power station. The signs are good.

Perhaps in 50 years' time when fusion power stations begin to produce electricity on a grand scale the era of fission reactors with their potential risks will seem like a bad dream. Until then they are a vital resource.

CHAPTER FIVE

The Ozone Hole

The discovery which raised the environmental consciousness of the public more than anything else in the 1980s was almost certainly that of a hole in the ozone layer. This discovery was made by Dr Joe Farman of the British Antarctic Survey. A thinning in the ozone layer was first noticed in 1982. By 1985 it was clear that the appearance of a 'hole' in the ozone layer was a regular springtime event. Here for the first time was proof positive that man's activities were having a real and potentially disastrous effect on our environment, in a place where few might have expected it – at the remote fringe of space many miles over the Antarctic.

It had been forecast. In 1974 Dr Mario Molina and Professor Sherwood Rowland of the University of California predicted a decline in global ozone if the production of what are known as chlorofluorocarbons, or CFCs, continued. Their conclusions were controversial and were challenged, by industry in particular. By 1985 there was no longer any doubt. The thin diffuse layer of ozone which protects the earth from the harmful ultraviolet rays of the sun was slowly being destroyed by man's foolhardiness. Curiously, readings from the American weather satellite Nimbus 7 had failed to show the hole in the ozone layer. It was only after Dr Farman had published his ground-based observations in the scientific journal *Nature* that scientists went back to the Nimbus readings and discovered that in fact the satellite had been registering no ozone layer over the Antarctic, but the computer which processed its findings had been discarding the readings as so unexpected that they could not be true. By the time the computer's readout was seen by a human scientist the evidence had already unwittingly been concealed. Firm evidence that CFCs were responsible for the damage came to light in 1987 following the Airborne Antarctic Ozone Experiment organised by NASA.

The hole in the ozone layer appears each spring and is as large as the United States and as deep as Mount Everest. In the spring of 1989 measurements showed that the ozone hole was as big as ever, and it began to break up earlier than expected spilling ozone-depleted air over the tip of South America and the Falkland Islands. And recent results show that it is not just over the Antarctic that ozone is being depleted. Since 1970 the ozone layer over the Northern hemisphere, where most people live, has shrunk by 4 per cent in winter and by 1 per cent in summer. It is estimated that if the emissions of CFCs maintain the same growth rate then the ozone layer will have been depleted by 20 per cent within the lifetime of today's children.

WHAT IS OZONE AND HOW IS THE HOLE FORMED?

Ozone is a form of oxygen. A molecule of oxygen in the air we breathe consists of two atoms of oxygen joined together – hence its formula: O_2. In the thin atmosphere between 15 and 50 kilometres up, the ultraviolet light breaks up the oxygen molecules into single atoms which then reassemble in threes giving rise to ozone or O_3, which is in fact a poisonous chemical. Ozone is made mostly over the equator where the sun's rays are strongest and is then transported around the rest of the world by the winds in the stratosphere. CFCs, meanwhile, diffuse up into the stratosphere where they are exposed to the sun's ultraviolet rays and break up forming hydrochloric acid gas. At the poles, on the surface of ice crystals formed in the extremely cold clouds, the hydrochloric acid is converted to chlorine molecules. When the sun rises in springtime, at the end of the long polar night, the chlorine molecules are converted into chlorine atoms which attack the ozone layer. The chlorine reacts with the ozone reforming it into oxygen. The chlorine survives these chemical reactions and repeats them with other ozone molecules. One chlorine atom can break down as many as 10 000 molecules of ozone in this way.

But why does it happen particularly at the South Pole? The key is the temperature. It is thought that circulation patterns in the atmosphere cause a vortex at the Polar regions which isolates air over the Arctic and the Antarctic. The Antarctic is much colder than the Arctic, which is why the problem is so acute over the South Pole. In the winter the temperature in the Antarctic can reach −80°C and it is at this temperature that the ice crystals form in the clouds.

WHAT ARE CFCs?

Chlorofluorocarbons were invented in 1930. There are several kinds, of which the two most widely used are CFC 11 and CFC 12. It is their very stability which makes them so useful to man but which also makes them such a menace in the atmosphere. CFC 11 lasts for an average of 74 years in the atmosphere and has been used extensively as a propellant in aerosols and as a coolant in refrigeration. CFC 12 has similar uses and lasts 111 years in the atmosphere. While the use of CFCs in aerosols has declined markedly, it is estimated, for example, that there are still some 30 000 tons of CFCs 11 and 12 in the cooling systems of existing refrigerators and air conditioning systems. CFCs have also been used as solvents and in the manufacture of foams for food packaging and insulation. (In fact there are more CFCs in the insulating material of some fridges than in the cooling systems themselves.) CFCs are currently increasing in the atmosphere by 6 per cent a year, though with recent decisions to limit their production the rate of increase should eventually reduce. Much more stringent decisions are required, however, to bring the concentration down again.

CFCs are not alone in harming the ozone layer. Another group of gases, called **halons**, used in fire fighting, also find their way to the stratosphere. They break down in ultraviolet light releasing bromine – which is many times stronger than chlorine as an ozone depleting gas (though the quantity of halons going into the atmosphere is far smaller than the amount of CFCs). Production of CFCs and halons in selected years since 1935 is given in Table 3. Production in 1989 is estimated to have been one million tons. Foam blowing, aerosols, refrigeration and solvent use have each contributed about 20–5 per cent of total world use of CFCs. Ironically, despite all the concern – and five years after the ozone hole was discovered – the total amount of CFCs produced in 1989 world wide was the highest ever.

In October 1988 Britain's Stratospheric Ozone Review Group, set up by the Department of the Environment and the Meteorological Office to look at what was known about the ozone problem in the stratosphere, published a report which demonstrated that over the Antarctic the depletion of ozone in the spring of 1987 was more severe and had lasted longer than previously. The amount of ozone above any particular point had decreased by 40 per cent compared with its pre-1975 thickness.

The report also forecast that the ozone hole would continue to appear every year until the chlorine levels in the stratosphere fall to levels which were present in the mid-1970s. The problem is that even if no more man-made CFCs or halons are released into the atmosphere it will be many decades before there is a return to 'safe' levels, because they are such stable compounds. As has already been said, unless drastic action is taken to eliminate CFC release, it is anticipated that chlorine

TABLE 3 GLOBAL EMISSIONS OF SELECTED HALOCARBONS

(millions of kilogrammes per year)

Year	*CFC–11*	*CFC–12*	*CFC–22*	*CFC–113*	*Halon 1211*
1935	0.0	0.3	0.0	0.0	0.0
1950	5.6	28.8	0.1	0.0	0.0
1955	21.5	46.9	0.6	0.0	0.0
1960	38.2	88.5	2.6	2.5	0.0
1970	191.2	300.5	18.3	17.0	0.0
1980	264.5	388.6	71.9	97.3	0.0
1985	298.0	438.0	81.2	138.5	2.6

Source: *Environmental Data Report 1989/90* (United Nations Environment Programme). © UNEP

levels in the stratosphere will increase over the next 50 years because of the long life of CFCs already in the environment.

Recently there has been evidence that the Arctic is similarly threatened. In the winter of 1989 an expedition in high flying aircraft over the Arctic confirmed the gloomiest predictions. Chlorine monoxide, which is one of the chemicals formed by the breakdown of CFCs, was found in concentrations 50 times greater than normal. Though the scientists did not detect any reduction in ozone at that time they said the atmosphere was primed for ozone destruction as springtime approached.

WHY IS OZONE DEPLETION SO BAD?

There are many theoretical reasons why it is dangerous to destroy the ozone layer. Ozone absorbs the ultraviolet radiation from the sun so without ozone the amount of ultaviolet radiation reaching the surface of the earth will increase. People who are exposed to increased ultraviolet radiation are more likely to develop **skin cancer** and **cataracts**. In fact skin cancer is already increasing around the world with an estimated 300 000 cases a year in the United States alone. Ultraviolet radiation also impairs **the immune system**, making people more prone to infections. It is estimated that a 10 per cent loss of ozone in the stratosphere – something which could well happen in the next forty years if CFCs continue to increase at the present rate – could cause 1.5 million extra deaths from skin cancer and 5 million extra cataracts in the United States alone, though these figures have been disputed.

Increased UV radiation may also seriously upset the **balance of nature in the oceans**. Some forms of ultraviolet radiation can penetrate some way beneath the surface of the sea harming plankton and the larva of fish.

Ultraviolet light could change plant life and agriculture. Two-thirds of some 300 crops and other plant species tested for their tolerance of ultraviolet light have been found to be sensitive to it.

More ultraviolet light reaching earth would actually increase the amount of ozone formed here on earth, which would, in turn, increase the amount of photochemical smog at ground level. In addition, man-made materials like plastics are sensitive to ultraviolet radiation and are made brittle.

A report by the United Nations Environment Programme warned that if the ozone layer were to disappear 'the sun's ultraviolet light would sterilise the surface of the globe, annihilating all terrestrial life'. It described the ozone layer as 'a life-giving layer of poison . . . unique to our planet'.

ACTION TAKEN TO REDUCE CFCs

Following initial concern in the mid-seventies about the ozone layer, the United Nations Environment Programme, or UNEP as it is known, set up a committee in 1977 to review the research and give forecasts of what they thought might happen. This eventually led in 1985 to the **Vienna Convention for the Protection of the Ozone Layer**, which had been negotiated under the control of UNEP. The Convention set out the framework for cooperation in research, observations and the exchange of information, but made no regulatory controls. It came into force in 1988.

But some countries did not wait for controls to be imposed on them by international agreement – they took unilateral action. The United States, Canada, Norway and Sweden banned the non-essential use of CFCs in aerosols in the late 1970s and early 1980s. In 1980 the European Community agreed to put a ceiling on the production of CFCs 11 and 12 (the most common CFCs) and to cut their use in aerosols by 30 per cent. The Soviet Union has said it intends to use non-CFC aerosol propellants by 1993. In 1988 Sweden became the first country to legislate a phase-out of CFCs so they will be virtually eliminated by 1994.

The suggestion that industry should play its part in cutting its use of CFCs has had a mixed reception. Some branches of industry have been persuaded to take action while others have shown what can only be described as a blatant disregard for environmental concerns. For example, a recent survey of the microelectronics industry in Britain showed that around 65 per cent of companies questioned had no intention of phasing out CFCs.

Perhaps the most significant achievement has been the Montreal Protocol to the Vienna Convention, which was agreed in September 1987 and which came into force on 1 January 1989 at which time it had been signed by more than 40 countries and ratified by 30 of them, plus the European Community. It has been described by the Director of the United Nations Environment Programme, Dr Mostafa Tolba, as 'the first truly global agreement on the protection of the environment'.

The Montreal Protocol said that production of CFCs must be frozen at 1986 levels by 1990. By 1994 production must have been reduced to 80 per cent of the 1986 levels, and by 1999, to 50 per cent. On the subject of halons, the Protocol states that production and consumption must be frozen at 1986 levels from 1992.

The Protocol recognises that special provision should be made for developing countries and therefore countries like India are allowed to increase their use of CFCs up to 0.3 kg per head per year (though this figure is unlikely to be reached). Those who signed the protocol also agreed to review its provisions every four years.

The Montreal Protocol paved the way for drastic action to cut CFC production world wide. But since it was signed many experts have been asking whether even tighter controls are needed. In 1988 the UK Stratospheric Ozone Review Group predicted that even just to stabilise chlorine levels, large cuts in CFC emissions would be needed – CFC 12 emissions would have to be cut by 85 per cent. The Review Group stressed that the Montreal Protocol did not go far enough claiming that the only way to prevent further depletion and to allow the atmosphere time to recover was to phase out production of the major man-made carriers of chlorine and bromine to the stratosphere. This led the British Government and several European Governments to call for reductions of CFC production of at least 85 per cent by the turn of the century and the United States to call for the complete phasing out of CFCs and halons.

Britain had been one of the more sceptical of the developed

countries before the Montreal Protocol had been negotiated, but signalled its conversion to the cause of protecting the ozone layer by convening an international meeting on ozone in London in March 1989. In his speech to the conference dinner, the Prince of Wales said that Britain could be justifiably proud that it would meet the Montreal Protocol target, of a 50 per cent reduction in CFC use by 1999, ten years ahead of this deadline, and he pointed out that the achievement had been made possible 'by the thousands of ordinary consumers and environmentalists whose concerted pressure persuaded the aerosol manufacturers to phase out their use of ozone-depleting CFCs by the end of 1989'.

One of the main issues the conference addressed was the difficulty foreseen in persuading the developing countries to forego the use of CFCs in their industries. Quite understandably, countries like China and India are at the beginning of a major growth in the provision of household goods like refrigerators and resent any implication that they should forego the improved standard of living such goods afford. The annual consumption of CFCs per head of population in developed countries is more than 1 kg, compared with an annual consumption in developing countries of less than 0.005 kg.

While recognising the problem of increasing CFCs, several of the representatives of the developing world at the meeting made the point that it was the developed world which caused the problem and therefore the developed world should bear the cost of recovering the situation. Their view, quite reasonably, was that if developing countries are to be expected to forego the use of CFCs then the extra cost of using substitutes should be borne by the industrialised nations. The Indian Environment Minister reminded the conference of what he called the excellent principle of 'the polluter pays', which the developed world had adopted. But whatever the developed world does it cannot solve the problem on its own. If the industrialised countries cut back their production of CFCs by 50 per cent, as agreed by the treaty, it would need just four of the largest developing countries to produce CFCs up to the allowed treaty limit of 0.3 kg per person for global CFC production to increase by 50 per cent!

However, despite the fact that the Protocol galvanised so many countries into doing something about CFCs, it is plain that the problem was far from solved. The Protocol did not cover all ozone depleting gases, for example methyl chloroform, which contributes 5 per cent to ozone depletion, and carbon tetrachloride, which contributes 8 per cent, were not mentioned. And it did not control those CFCs which are less of a threat to the ozone layer because they break down more easily in the lower atmosphere. The American Environmental Protection Agency estimated that these uncontrolled compounds account for about 40 per cent of the projected growth in stratospheric chlorine levels by 2075.

With the increased realisation the the Montreal Protocol did not go far enough it was hardly surprising that the review conference, held in London in June 1990, succeeded in tightening up the controls on ozone-depleting chemicals. Nearly one hundred countries attended the conference and by the end had reached a remarkable degree of agreement. They agreed, for example, to a phase out of CFCs by the year 2000. About a dozen countries argued strongly for an even quicker phase out – by the year 1997 and, though they could not persuade the rest, including CFC-producing countries like America and Japan and the USSR, to follow their example, pledged that they themselves would complete the phase out by then.

Other ozone-depleting chemicals were included in the Protocol. The meeting agreed that production of halons would be halved by 1995 and phased out by the year 2000. The Protocol was also extended to include two other important ozone-depleting chemicals: carbon tetrachloride and the metal cleaner, methyl chloroform.

Some of the substitute chemicals being developed by industry to take the place of CFCs – the so-called HCFCs – are not entirely ozone-friendly and these and others can also contribute to global warming. The countries approved a resolution which should mean they will be phased out by 2040 at the latest.

Once again, the Chinese and the Indians argued strongly for the 'polluter pays' principle. India, this time represented by the charismatic Mrs Maneka Ghandi, made a powerful case for the developing world to be given the technical know-how on CFC substitutes so more developing countries could sign the Montreal Protocol without fear of jeopardising their development. Meetings with representatives of industry succeeded in persuading her that industry would do all it could to help. And, after further late-night sessions with officials, India won the concession that unless it was given the technology it would not have to abide by the Protocol.

The other main part of the new agreement was that a fund worth $240 million for the first three years should be set up with new money to help the Third World in its transfer to CFC substitutes. Once these details had been tied up, India and China indicated that they too would join the Protocol. So within the space of three short years a unique international agreement was forged to help save the ozone layer – though even this did not satisfy some environmentalists who pointed out that even tougher action would be needed if the ozone hole was to be repaired in the next century.

WHAT ARE THE ALTERNATIVES?

One of the first major reductions in CFC use was in the aerosol industry. In many cases CFCs were replaced by hydrocarbons. The industry has borne the brunt of public concern about CFCs. Even though now in Britain only an estimated 10 per cent of aerosols still contain CFCs, until very recently some people (including the Prince of Wales) still referred to aerosols in a pejorative way. The substitution of CFCs with hydrocarbons has not been entirely beneficial because hydrocarbons are themselves greenhouse gases. Many environmentalists, including organisations like Friends of the Earth, have been urging people to make do with pump action compressed air aerosol sprays.

With foam production and food packaging new chemicals are being used instead of CFCs, and in some cases there is a return to ordinary cardboard containers. In buildings, insulation fibreglass can be used instead of polystyrene foam. And the use of CFCs as solvents in cleaning, particularly in the electronics industry, can be easily avoided.

The chemical industry world wide is actively engaged in the production of alternatives to CFCs 11 and 12. The new substances include HFC-134a, HCFC-123 and HCFC 141b.

HFC-134a is being manufactured by the big American chemical giant Du Pont, and ICI, which was one of Europe's largest CFC producers. It will be used in refrigerators and air conditioners provided it passes all the safety tests. Altogether some 14 countries from the United States, Europe, Japan and South Korea have joined forces to test CFC substitutes to speed up the process of substitution. In America companies which account for 85 per cent of CFC production have said they intend to phase out the manufacture of all CFCs covered by the Montreal Protocol, no doubt in response to the enormous public pressure which has built up against CFCs.

Even before substitutes are introduced there is scope for saving on the production and emissions of CFCs by better housekeeping and by recycling. CFCs used as cleaning agents in the computer industry can be recycled. In West Germany an IBM plant has installed a recycling system which recovers up to 90 per cent of the CFCs used. Most of the CFC 11 used in the process of foam blowing can be recovered. And the CFC used as a refrigerant can be recovered when a refrigerator is discarded.

The world has at last woken up to the seriousness of the damage being done to the ozone layer by CFCs and action is being taken to combat it. Whether it will be enough depends on how the action is accelerated in the next few years. Perhaps, if hard evidence accrues that cases of skin cancer in particular are increasing dramatically in areas under the ozone hole, then the phase-out of CFCs will be speeded up. Whatever is done to counteract the problem, scientists are fairly confident it could take a hundred years or more before the damage already put in train in the atmosphere is repaired.

CHAPTER SIX

Air Pollution and Acid Rain

The hole in the ozone layer is not something which most people can witness for themselves and so they rely on experts to tell them it is important. But one aspect of the pollution of the atmosphere, of which people are all too aware, is the smoke and gases that come from power stations, factories and car exhausts. Air pollution is a menace which affects us all.

World wide, more than a billion people – a fifth of the world's population – live in communities which do not meet the basic air quality standards set by the World Health Organisation. In Bombay, simply breathing is equivalent to smoking ten cigarettes a day. In the United States it is said that air pollution causes as many as 50 000 deaths a year. Paris, Madrid, Rio de Janeiro and Milan are among the cities most polluted with sulphur dioxide.

Technological fixes for some of the causes of pollution have helped modify the problems in places. The judicious use of taxes continues to encourage energy saving and less polluting fuels. But the use of technology and fiscal measures is primarily a feature of Western democracies. They have still to make an impact in most of Eastern Europe and the developing world. One of the worst countries for air pollution is Poland. There air pollution from the giant Nova Huta steel works makes the air for miles around taste like tar.

Indeed there are serious environmental pollution problems in many Central and Eastern European countries caused mainly by the burning of brown coal by heavy industries. The discharges from these coal burning plants go into the air and water without treatment. Technology for environmental protection is not advanced and legislation, where it exists, is either poorly operated or not operated at all. The most serious pollution occurs in Poland, East Germany and Czechoslovakia.

Air pollution is not only a problem of industrialised countries. In the Third World, air pollution of a different sort is causing ill-health. In some countries – in Africa for example – smoke from indoor cooking fires causes severe lung disease in infants. Reports from Burkina Faso and Zambia suggest that respiratory illness ranks among the top two or three causes of illness. Pregnant women exposed to indoor smoke are at risk of producing small babies in the same way that women who smoke cigarettes tend to have lighter than average offspring. Carbon monoxide in air reduces the blood's ability to carry oxygen and this is liable to pose a risk for people suffering from heart disease. Another component, the oxides of nitrogen, are powerful lung irritants and can reduce resistance to infections like flu.

One of the most important consequences of air pollution is the production of acid rain.

WHAT IS ACID RAIN?

The two types of gases which cause the most problems are sulphur dioxide (SO_2) and the oxides of nitrogen. A proportion of these gases is absorbed directly by plants and materials but the remainder stays in the atmosphere where it may be oxidised to acids, particularly sulphuric acid and nitric acid.

Acidity is measured by a unit known as pH. If water is said to have a pH of 7.0 it means it is neutral – neither acidic nor alkaline. A pH higher than 7 implies the water is alkaline and a pH less than 7 indicates it is acidic. The lower the figure the more acidic is the water. The pH scale is logarithmic, which means that a decrease of one on the pH scale represents a ten-fold increase in acidity. Normal rain is slightly acidic, even if it has not been contaminated with any gases produced by man's activities, because it contains some dissolved carbon dioxide from the air and some naturally occurring sulphuric acid. Uncontaminated rain probably has a pH of about 5.0 but acid rain due to man-made pollutants can have a pH much lower than this.

Acidity can reach the earth either in the wet form, as acid rain, or in a dry form either as a gas or as a gas which has been absorbed by dust. Damage can be caused also by acidity trapped in snow and in fog. In the greater Los Angeles area in the United States the pH of fog has on occasion fallen as low as 2 – about the acidity of lemon juice. Similar values have been reported for clouds in the northern Appalachians. Acid rain is less acid. In the north eastern United States, for example, it has an average pH of 4.2. The term acid rain is often used as a generic term to cover all forms of acid deposition.

EFFECTS ON LAKES AND RIVERS

It was in the late 1960s and early 1970s that scientists first became worried about acid rain in the environment. There were reports of increased acidity in rain in the upland areas of southern Scandinavia as well as in the northeast United States. There were suspicions that the acidity was affecting water quality, freshwater fisheries and trees. Recent tests on many lakes in the United States have shown that their acidity has increased in the last few decades. In six of eleven Adirondack lakes, for example, the pH has fallen since the 1930s to 5.2, probably as a result of the highly acidic rain in western New York (pH 4.1) It was recognised that the acidity was probably coming not only from local sources but from a long way away. There were suspicions too that the effect was not caused by a sudden change, but had been building up over the last 100 years.

In Britain a study in 1978–80 showed that in northwest Scotland rain had a pH of 4.7, in northeast England and southeast Scotland it was 4.2. The most acidic rain occurred in southern England where it measured 4.1. When the volume of rain is taken into account, the largest inputs of acid rain in Britain are parts of Cumbria, in the southern uplands and west central highlands of Scotland and the Welsh mountains. The amount of acid rain is of the same order as that which has affected Scandinavian and American lakes and damaged their fish stocks. The Royal Commission on Environmental Pollution in its 1984 report commented that many water authorities in Britain had expressed concern to the Commission about acidification of lakes and the effect it was having on the biological activity. In particular, in the spring, when snow melts, some lakes can get a pulse of acidity and this can be particularly devastating to fish stocks. pH values as low as 3.0–4.0 can occur. There is recent evidence showing that lakes in southwest Scotland have become acidified over the last 150 years as a result of air pollution.

EVIDENCE OF ACIDIFIED LAKES IN CERTAIN COUNTRIES

Country	*Evidence*
Canada:	In the east of the country 150 000 lakes (1 in 7) show evidence of some biological damage. More than 14 000 lakes are strongly acidified.
Norway:	In 13 000 square kilometres of water there are no fish. In a further 20 000 square kilometres fish show signs of being affected.
Sweden:	14 000 lakes unable to support sensitive aquatic life and 2200 nearly lifeless.
United Kingdom:	Some acidified lakes in southwest Scotland, in west Wales and in the Lake District.
United States:	In 1984 the Environmental Protection Agency found 552 lakes strongly acidic and 964 marginally acidic.

Source: *State of the World 1990, A Worldwatch Institute Report on Progress Toward a Sustainable Society* © Worldwatch Inst.

It is not just the acid rain itself which, by increasing acidity in lakes, may be having a serious effect on fish stocks. The acidity can also interact with the soil through which it drains, releasing toxic metals such as aluminium. Changes in agriculture and forestry can also have a marked effect on the acidity of run-off water.

One of the most controversial aspects of acid rain is that it crosses national frontiers. Acidity produced in the atmosphere in the industrial centres of Britain, for example, can be carried high in the atmosphere across the North Sea and affect lakes in Scandinavia. And Canada has complained bitterly about acid rain originating in the United States. This resentment is, of course, entirely understandable.

Lakes which have become so acidic that the fish have died can be temporarily rescued by 'liming'. Neutralising lime can be dumped into the lake from aircraft or it can be scattered on the surrounding hills to be carried into the lake by water running off the land. In an experiment in Loch Fleet in Scotland the CEGB carried out a programme of liming and managed to return the loch to a state in which it was once more able to support fish.

EFFECTS ON FORESTS

And it is not just lakes and surface waters which are suffering from acidity: many of the world's forests are also being seriously affected, though the blame is not so easily laid at the door of power stations and industry with their sulphur dioxide emissions. The evidence now building up is that it is emissions of the oxides of nitrogen which may be as important, and they come not only from power stations and industry but from cars and other vehicles. The most dramatic effects on trees have been seen in the forests of Germany – as illustrated in Table 4 – but forests elsewhere have suffered, particularly in the eastern United States. Here the most serious damage has occurred in the coniferous forests in the mountains at heights above 850 metres. Areas badly affected include the Adirondacks, the Green Mountains in Vermont and the White Mountains in New Hampshire, where about a quarter of the red spruce have died in the past 25 years and those which survive are growing less vigorously. What makes it difficult to identify air pollution as the culprit is the fact that the forests are often very far away from any obvious sources of pollution which might be considered strong enough to poison trees.

The decline of forests can, it is suggested, occur over a period of some 50 years, during which there is little or no detectable damage to trees. If that is true it means that acid rain may now be affecting many countries where it has yet to be recognised. Perhaps even the delicate ecosystems of the Arctic and Antarctic are being damaged.

TABLE 4 FOREST DAMAGE IN SELECTED EUROPEAN COUNTRIES: 1988

Country	*Total forest area*	*Estimated area damaged*	*Share of total*
	(thousand hectares)		*(per cent)*
Czechoslovakia	4578	3250	71
Greece	2034	1302	64
United Kingdom	2200	1408	64
West Germany	7360	3827	52
Norway	5925	2963	50
Denmark	466	228	49
East Germany	2955	1300	44
Switzerland	1186	510	43

Source: *State of the World 1990, A Worldwatch Institute Report on Progress Toward a Sustainable Society* © Worldwatch Inst.

EFFECTS ON VEGETATION

And of course the damage does not stop at trees: the organisms which have suffered more than almost any other as a result of air pollution are lichens. They are Britain's most sensitive indicator of air pollution. There are 1500 species of lichens in Britain but now areas rich in these tiny plants are confined to an L-shaped zone from south Kent westwards to Cornwall and north up the west coast through Wales to western Scotland. Here clean air from the Atlantic blows in and air pollution and acid rain do not penetrate.

The impact acid air pollution can have depends on various complicating factors including the weather. Even where sulphur dioxide is present only in low concentrations, in high upland regions where weather is harsher the impact of acid rain can be greater. This is because the vegetation on hills is exposed to frequent mists and clouds. Clouds may have between two and eight times the concentration of acid in them than acid rain. Plants on mountainsides may also be subjected to pollution from ozone and a substance called PAN (peroxyacetyl nitrate) formed by the action of sunlight on polluted air, and a component of urban smog. The fact that forests are good at collecting polluted cloudwater from the atmosphere may also be one reason why lakes in upland forested areas seem more acidic. Because of this there is a fear that as young forests mature they will actually increase the extent of acidification in lakes and surface waters in Scotland and Wales even though sulphur dioxide emissions into the atmosphere will have come down.

EFFECTS ON BUILDINGS

Quite apart from the serious effects acidity in the atmosphere is having on lakes, rivers and vegetation it is also having a most harmful effect on buildings both ancient and modern. One only has to look at the crumbling stonework of historic buildings like Westminster Abbey, now being magnificently restored, to recognise the seriousness of the problem. The stonework becomes friable and bits can be broken off quite easily. The cost of restoration and repair is so huge that it is comparable to the cost of installing clean-up equipment at power stations – a powerful additional argument for reducing emissions.

In Eastern Europe the problems are far worse than in the West. Levels of sulphur dioxide in the air in the Polish city of Krakow reaches 90–100 microgrammes per cubic metre and can sometimes go as high as 200. By way of comparison, in the polluted parts of the United States it reaches only about 10 microgrammes per cubic metre with 2 parts per cubic metre being about the normal level.

HOW CAN ACID RAIN BE PREVENTED?

The most effective and obvious way to prevent acid rain is to stop the pollutants from being emitted in the first place. There are three main sources of the polluting gases – power stations, industrial plants which burn fossil fuels, especially coal, and vehicles which use petrol or diesel. In Britain nearly 70 per cent of emissions of sulphur dioxide and some 46 per cent of nitrogen oxide emissions come from power stations. In the mid-eighties the CEGB was suggesting that reducing emissions of sulphur dioxide from power stations would have little effect on acid rain because what was critical for transforming sulphur dioxide and the oxides of nitrogen to acid rain was the presence of hydrocarbons and photochemical oxidants. This view, however, has not been popular with some. In America, for example, the National Academy of Sciences has concluded that there is a direct connection between emissions of sulphur dioxide and nitrogen oxides and acid rain over the eastern half of the United States.

In Britain fossil fuel power stations emit 3.25 million tonnes of sulphur dioxide a year. In 1984 the CEGB estimated that with modest economic growth and the increase in nuclear power generation which was then expected, the amount of sulphur dioxide put out by the year 2000 would be somewhat similar. The CEGB estimated it would take ten years to fit their ten largest coal-fired power stations, which account for about 50 per cent of the CEGB's total emissions of sulphur dioxide, with flue gas desulphurisation plants. The capital cost they estimated would be £1500 million and they said the plants would increase electricity bills by 6 per cent. The CEGB was clearly not keen at that time to build the plants, describing them as 'expensive, wasteful of energy, and by their consumption of raw materials and waste production damaging to sections of the environment other than air'. They are in reality large chemical plants built alongside the power stations and almost as big.

In 1985, in a briefing document for MPs, the CEGB put up a strong and detailed case for not rushing into major policy decisions costing billions of pounds until there was more scientific evidence for what was needed. The document stressed that determining the cause of damage to forests in Germany and lakes in Sweden and Norway was a highly complex problem.

More recently, however, the scientific consensus is that sulphur emissions are, in fact, responsible for lake acidification and the CEGB position has changed to one of broad acceptance.

Ironically, it was in Britain that flue gas desulphurisation was invented as long ago as the 1880s. After the Second World War there was talk of fitting all new power stations with the equipment. What made it seem likely was the terrible London smog of 1952 which killed 4000 people and which paved the way for the Clean Air Act of 1956 which began the process of getting rid of smoke from towns and cities. In 1954 the Beaver Report on the smog urged that both smoke and sulphur dioxide should be removed from power station emissions and stressed that sulphur dioxide 'was one of the most harmful pollutants'. But in the end only three power stations were fitted with the desulphurisation equipment and the Government decided, instead, that in future no new power stations would be sited in towns. The smoke was mostly removed from power stations but the sulphur dioxide was left.

However, partly in response to improved scientific information and also because of public pressure, in September 1986 the CEGB appeared more enthusiastic about flue gas desulphurisation and announced that it would fit the plants to three coal-fired power stations to reduce sulphur emissions by 10 per cent by the year 2000. One reason for this decision was a new estimate by the Department of Energy that unless something was done emissions might actually rise by 30 per cent by the year 2000 – taking total emission levels back to what they were in 1980. The Government had already pledged that emissions would be reduced by 30 per cent by the end of the century.

Despite the commitment to fit flue gas desulphurisation plants it took a long time for the first contract to be agreed.. It was not until February 1989 that the Government announced that the CEGB had signed a contract to fit the first plant – to the Drax power station in Yorkshire in what was believed to be the world's biggest retrofitting of such a plant to an existing power station. The Government announced at the time that to comply with an EC directive they were determined to cut sulphur emissions from existing plants by 20 per cent of the levels recorded in 1980 by 1993, by 40 per cent by 1998 and by 60 per cent by 2003.

The decision to fit three large power stations with desulphurisation equipment might appear to be a step – even if a small step – in the right direction, but it did not satisfy the Swedish Environmental Protection Board which said that 90 per cent reductions were necessary. Britain is planning such reductions by the year 2020 but Scandinavian countries are getting impatient. They claim that seven major salmon fisheries in Norway are now devoid of any fish and up to 6000 Swedish lakes have been affected.

A further concern is that flue gas desulphurisation does little to remove nitrogen oxides, so to reduce their emission what are called low NOx burners are to be fitted to 12 major coal-fired power stations before the end of the century. It is claimed that low NOx burners could cut the emissions of nitrogen oxides from power stations by 30 per cent of 1980 levels by the year 2000.

As we have already seen, with the increasing concern about the levels of carbon dioxide entering the atmosphere from coal-burning power stations there are moves to increase the use of natural gas. Burning natural gas, however, will still produce nitrogen oxides. Indeed, any combustion process, even burning hydrogen in air, will produce oxides of nitrogen by the direct combination of atmospheric nitrogen and oxygen in any flame at high temperatures.

Another source of anxiety is the increasing production of ammonia which aggravates the acid rain problem. The main source of ammonia is agriculture. It evaporates from fertiliser in arable areas and is also given off by animal waste. In the Netherlands there is evidence of major environmental problems as a result of ammonia emissions from intensive livestock rearing. There are major changes occurring in vegetation type – from heather to grassland, the forests are in decline and there is nitrate production in groundwater. There is also evidence that ammonium ions in rain can be transported over long distances and deposited in areas which are not used to getting large amounts of nitrogen, such as upland forests. This may result in the decline of forests because of nutrient imbalances even though, paradoxically, the additional nitrogen from the ammonia initially causes the trees to grow better.

There have been calls from organisations like Wildlife Link – the liaison body for all the major voluntary organisations in Britain concerned with the protection of wildlife – for an urgent investigation into ways of reducing ammonia emissions. In the Netherlands, for example, it has been suggested that increases of stocks of pigs, cattle and poultry should be curtailed, that there should be controls on the use of manure and that manure in storage should be covered to prevent ammonia evaporating. It has even been suggested that manure should be injected or ploughed into the ground rather than spread on top, to reduce the evaporation of ammonia.

Vehicles are another major source of emissions which cause acid rain. Engines are becoming more efficient and in theory it should be possible to design an engine which mixes just enough air with the petrol to achieve complete combustion, with no waste of petrol and no emissions of pollutants except the products of complete combustion (water vapour, carbon dioxide and nitrogen). That, of course, does not happen. Instead, various pollutants are produced including nitrogen oxides, carbon monoxide, hydrocarbons and particulates. Nitrogen oxides from cars (both petrol and diesel) contribute about 40 per cent of the total emissions in Britain – a similar amount to that produced by power stations.

INTERNATIONAL ACTION TO CONTROL EMISSIONS

International action to control air pollution began in earnest in 1983 with the drawing up of what is rather long windedly called The United Nations Economic Commission for Europe Convention on Long Range Transboundary Air Pollution. Thirty-one parties contracted to the Convention, including the European Community, and they tackled a number of issues.

By 1986, 21 countries had signed **Sulphur Dioxide Protocol** under which they were committed to reducing sulphur dioxide emissions by at least 30 per cent of 1980 levels by 1993. This is the so-called '30 per cent' club. But by 1986 two of the world's four largest sulphur emitters, the United States and Britain, had still not signed it. (Britain claimed it had already done more than most to reduce its SO_2 emissions before 1980 and so to expect more from Britain was unreasonable.) And 14 of the 21 who had signed it had still not ratified it (it needs 16 ratifications to bring it into force). The slowness with which the Protocol has evolved and the reluctance of certain countries to sign it, led the World Wide Fund for Nature to accuse those countries of showing 'a deplorably low level of commitment' to the Convention.

Then, in 1987, negotiations towards an **Oxides of Nitrogen Protocol** began. This contains two basic obligations: a) an initial freeze on total national nitrogen oxide emissions so that by the end of 1994 they do not exceed those of 1987, and b) a commitment to developing a scientific basis for revising the Protocol so that some form of agreement to reduce emissions is achieved, beginning no later than 1996.

Various initiatives to reduce emissions have come from other sources in addition to the United Nations. For example, in November 1988 European Ministers adopted the European Community 'Large Combustion Plants Directive'. This committed countries to achieving major reductions in emissions of the sulphur dioxide and oxides of nitrogen. It applies to large combustion plants over 50 MW thermal input and includes power stations and other industrial power plants. Member countries have to reduce emissions of sulphur dioxide from existing large plants by 60 per cent of 1980 levels by the year 2003. Emissions of nitrogen oxides have to be reduced by 30 per cent of 1980 levels by 1998.

As far as sulphur is concerned, in 1980 Britain emitted 4.7 million tonnes of sulphur dioxide so the agreement means that figure would have to come down to 1.9 million tonnes by 2003 – a very difficult goal to reach, especially since the Government in November 1989 abandoned its plans to build three new PWR nuclear power stations. At the time CEGB said it would meet the new standards by approximately doubling its programme for fitting flue gas desulphurisation equipment to power stations (from 6000 to 12 000 megawatts). Some experts believe that to achieve the stated goals, practically all Britain's coal-fired power stations would need to be fitted with the clean-up equipment at a total cost of £3 billion. But there are other ways of meeting the commitment to cut sulphur emissions, including using more gas-burning power stations and importing less sulphurous coal. A suggestion in February 1990 that this was indeed how Britain planned to meet initial targets brought a howl of disapproval from environmentalists.

In the last 10 years Britain has gone a long way towards recognising its contribution to the international problem of acid rain and at last moves are being made to reduce Britain's emissions. Many would claim Britain has not gone far enough, and there is certainly scope for greater reduction.

CHAPTER SEVEN

The Infernal Combustion Engine

If there is one consumer product which is top of the popularity polls it is the car. Car ownership is rising throughout the world – and with it increased air pollution and increased congestion. While individual cars may be becoming less polluting, the sheer increase in the quantity of vehicles means the net effect is much greater air pollution.

There are nearly 400 million cars in the world today compared with about 50 million immediately after the war. They are rolling off production lines at the rate of 126 000 a day. Europe has the biggest car manufacturer in the world – Volkswagen. Its Wolfsburg factory in Germany produces a new vehicle every fourteen seconds. Across Europe car production now exceeds the birth rate.

At the turn of the century few envisaged the impact cars would make. In 1901 Mercedes Benz estimated that the world market for cars would be no more than 1 million. By the year 1915 that number had already been passed. Production has grown rapidly since, from fewer than 10 million vehicles a year in 1950 to nearly 30 million in 1973 and nearly 33 million in 1987. For years America dominated the world car market and it was only in the late sixties that car production elsewhere in the world exceeded that in the United States. By 1969 European car production exceeded that of both the United States and Canada. In the seventies Japan emerged as the most dynamic producer and is now responsible for 8 million cars a year. America, though, still accounts for a quarter of the world's car production and one-third of its car fleet – it is the largest single market, with Japan second.

Car ownership is still predominantly a characteristic of the industrialised countries. In 1985 fewer than 1 per cent of the Third World's population owned a car. While China and India account for 38 per cent of the world's population only 1 per cent of their people own a car. Car ownership is rising rapidly, however, and it is estimated that by the year 2000 there will be 4 million cars in China.

Cars are almost wholly reliant on the oil industry and consume an ever-increasing share of its products. Indeed in the United States 63 per cent of all the oil used goes to transport, compared with 44 per cent in Western Europe, 35 per cent in Japan and 49 per cent in developing countries. In the United States oil reserves are well on their way to depletion, output dropped in 1986 and 1987 and imports increased by 30 per cent. As oil supplies dwindle cars will become increasingly dependent on the Middle East for oil.

But it is not so much what goes into the car's fuel tank which is causing such soul-searching at the moment but what comes out of the exhaust pipe. Car emissions produce alarming effects and despite efforts to reduce polluting emissions there is little real hope of a dramatic change in the short term. One estimate from the University of California suggests that emissions may be causing as many as 30 000 deaths a year in the United States.

Petrol engines produce a host of pollutants including lead, carbon monoxide and carbon dioxide, oxides of nitrogen and of sulphur, particles of soot and unburned hydrocarbons. Figures from the Organisation for Economic Cooperation and Development suggest that the internal combustion engine is responsible for three-quarters of all the carbon monoxide in the air and nearly half the nitrogen oxides. World wide, vehicles are responsible for 17 per cent of carbon dioxide released from fossil fuels.

One of the most serious results of all this air pollution is the formation of **photochemical smog** a brown haze which hangs over many cities and which causes lung damage. The worst pollutant in the smog is ozone produced by the chemical reaction between the oxides of nitrogen and hydrocarbons in the presence of sunlight. Apart from damaging human health it also damages crops and is partly responsible for the damage to trees in forests. In Germany it is said that 60 per cent of the trees in the Black Forest are now damaged due to car pollution. And figures from around the world show the seriousness of the situation in many towns and cities.

Up to 75 million Americans are said to live in areas which fail to meet air quality standards for ozone, carbon monoxide and particulates. In Budapest the level of carbon dioxide is two and a half times Hungary's permissible level. In Athens smog is said to claim as many as six lives a day. And it is estimated that more than half the people who live in Calcutta suffer from respiratory disease caused by air pollution.

One pollutant which is relatively easy to deal with is **lead**. Lead is added to petrol to increase the octane rating, but it is not essential and many countries have taken action to prevent it being added to petrol following indications that even very low levels of lead may well be causing intellectual impairment in young children. After the United States and Japan banned the use of leaded petrol in new cars (in 1975 and 1976 respectively) lead emissions decreased substantially. In the United States the total annual lead emissions decreased by 94 per cent. Over the same period the lead level in the blood of the average American dropped by more than a third.

Europe lagged behind America and Japan in removing lead from petrol largely because of powerful lobbies from the motor industry. In Europe, Austria, the Netherlands, Scandinavia, Switzerland and West Germany led the way. In Britain, following a sustained lobby by the Campaign for Lead Free Air, by September 1989 unleaded petrol accounted for 25 per cent of petrol sales. The European Commission estimates that by the year 2000 unleaded petrol will account for 83 per cent of sales.

Removing the three other main pollutants from exhaust gases (nitrogen oxides, carbon monoxide and hydrocarbons) requires what is called a **three-way catalytic converter** which is fitted to the exhaust system. The exhaust gases pass through a metal honeycomb in which are embedded small quantities of various metals like platinum and paladium. These act as catalysts, causing a chemical reaction to take place but not being altered themselves in the process. The honeycomb structure is necessary to increase the surface area of the catalysts so that as much of the exhaust gas as possible comes into contact with them. An ordinary three-way catalytic converter is contained in a ceramic 'shoe box' about 30 cm long by 22 cm wide. If the honeycomb were laid out flat it would cover an area the size of a football pitch.

The platinum and palladium convert the unburnt hydrocarbons and carbon monoxide into carbon dioxide, and another metal, rhodium, converts the oxides of nitrogen into nitrogen and water. To be properly effective a car fitted with a catalytic converter also needs a monitor in the exhaust system to determine how much oxygen is coming through. If there is too little or too much there is a feedback mechanism by which the oxygen flow to the engine is adjusted. Catalytic converters are poisoned by lead and therefore cars with catalytic converters have to run on unleaded fuel.

In the United States since 1977 all American cars have been equipped with catalytic converters to reduce their emissions. The US Environmental Protection Agency estimates that since 1983 carbon monoxide emissions in the atmosphere have been reduced by 88 per cent; hydrocarbons by 90 per cent and nitrogen oxides by 50 per cent.

The United States and Japan have set the pace for cleaning up car emissions. In the sixties the average car produced 10 grammes of hydrocarbons for every mile travelled. Today emissions are controlled and cars put out no more than 0.41 grammes per mile. It is the same story for carbon monoxide. In the uncontrolled sixties cars were emitting 80 grammes of carbon monoxide per mile. Now emissions are down to 3.4 grammes per mile. Nitrogen oxides have come down from 4 to 1 gramme per mile. Other countries are adopting similar emission standards. The European Community Environment Ministers in June 1989 voted to bring in much tighter controls on emissions in the EC which will mean all cars produced after 1992 having to be fitted with three-way catalytic converters. European manufacturers had hoped they would only have to fit them to cars with engines bigger than 1.4 litres, using lean burn technology on the smaller cars. But those hopes were dashed when the ministers agreed to stricter emission levels than expected.

The emission controls in the United States may have led to a reduction in overall pollution but now rising car ownership has meant the problems are returning. Carbon monoxide emissions have stopped falling in the United States and have stabilised. In Europe the fitting of catalytic converters will cut pollution, but if car ownership increases, emissions will eventually rise again. It is a similar story with hydrocarbon and nitrogen oxide emissions. One study by Earth Resources Research has suggested that one way of cutting emissions in Britain would be to enforce speed limits on motorways and dual carriageways: this could increase fuel efficiency by 2.5 per cent. Cutting speed limits to 60 miles per hour would double the saving.

The fact that the pollution problem is far from solved in the United States is clear from measurements of ozone levels in cities. The average daily ozone concentration in cities decreased by 15 per cent between 1975 and 1981 but only by half as much since then. Fifty-nine American cities do not meet federal carbon monoxide standards and a further nine also do not meet official ozone standards.

Diesel emissions are another problem. Diesel engines are becoming more popular, yet they are more difficult to fit with catalytic converters. Diesels emit particulate matter which among other things is potentially carcinogenic and blackens buildings.

It is clear then that despite all the measures which have so far been implemented air pollution by cars is still a big problem. A Worldwatch Institute report on 'Rethinking the Role of the Automobile' quotes the Administrator of the US Environmental Protection Agency as saying 'the smog problem may well need to be dealt with by reducing the number of cars on the street and by telling people they can't drive nearly to the extent they have in the past'.

If it is not possible to cut down the use of cars very easily then there is an alternative method of controlling exhaust pollution and that is to use less polluting fuels. Alternatives will in any case have to be found one day because petrol itself is a finite resource.

ALCOHOL

One of the most successful alternative fuels is alcohol produced from sugar-cane. Brazil has a highly successful industry producing **ethanol** in this way. In 1986, blended with petrol, it provided half the country's vehicle fuel. Now almost a third of Brazil's cars can run on pure ethanol. But Brazil is something of a special case, with a relatively small number of cars and a useful crop surplus. Other countries may not be able to exploit the potential of ethanol so easily. Ethanol could be made from other feed stock crops, but if maize were to be used for this purpose in America, for example, it would take 40 per cent of the entire American harvest to provide sufficient fuel for just 10 per cent of its cars.

From an environmental point of view the production process of all alcohol fuels is not very energy efficient. When you take into account the amount of energy needed to cultivate the crop, including the use of farmers' vehicles and the production of the fertilisers that would be needed, the end result would consume more energy than it would produce, though new techniques might make the process more energy efficient.

Ethanol would also cost more than petrol. Oil would have to increase in price from the present $23 a barrel to between $40 and $67 a barrel before ethanol would become competitive.

Another alcohol, **methanol**, would be somewhat more competitive. But fluctuating prices from year to year make cost comparison very unreliable. In any case, prices for petrol take no account of dwindling supplies or of its environmental impact. California has been looking with increasing enthusiasm at methanol made from coal or natural gas as a way of combatting its air pollution problem. It hopes eventually that methanol will replace up to 30 per cent of petrol consumption. Methanol has its disadvantages, however. It may well cut emissions of hydrocarbons but it does not reduce carbon monoxide emissions, and cars burning methanol produce two to five times as much of the potential carcinogen formaldehyde compared with cars using petrol. Formaldehyde also encourages the formation of ozone. As far as the production of carbon dioxide is concerned methanol has no advantages. If it were made from coal, that process itself would double carbon dioxide emissions.

NATURAL GAS

Natural gas is another fuel which might be used either in compressed form or in liquefied form. In Italy, for example, there are 300 000 cars which run on compressed gas and many other countries are beginning to use it.

HYDROGEN

One fuel which seems devoid of most of the problems which arise with other fuels is hydrogen. It can be generated from a variety of things, incuding coal and oil and gas, but most promisingly of all from water through electrolysis (the process in which an electric current is used to split water into oxygen and hydrogen).

Hydrogen is about 45 per cent more efficient than petrol. Its use has yet to become a practical proposition with vehicle technology still in the prototype stage. It is also expensive to produce, but in theory it does have some key advantages, most notably the fact that it does not pollute. When hydrogen is burnt it produces nothing but water.

ELECTRICITY

Electric vehicles are more efficient and quieter than existing petrol-based engines but they don't have the range or performance. And if the electricity is produced in ordinary power stations then there would be no net saving on carbon dioxide emissions and more sulphur dioxide would be produced in total. Fuel cells which can make electricity directly from hydrogen may be useful one day.

MAKING CARS MORE EFFICIENT

One way to reduce fuel consumption is to make cars more efficient. In America, before the oil crisis of 1973, the average car travelled 13 miles to the gallon. By 1986 that had improved to 18 miles per gallon. New US cars now being produced can go 27 miles per gallon – still below European and Japanese cars which go 30 miles to the gallon on average.

One obvious way to make cars more efficient is to ensure that they carry more than one person: in the United States car pooling arrangements have been attempted but they have a long way to go to achieve their full potential. In the United States in 1984 the amount of energy used per passenger mile travelled was just as high as in 1971.

A car's efficiency can be improved by making it more aerodynamic (reducing a vehicle's drag by 10 per cent will save 5–6 per cent of its fuel on the open road and 2–3 per cent in city driving), making it lighter (a 10 per cent weight reduction produces a 6 per cent gain in fuel economy), reducing the resistance of the tyres on the road, reducing the losses of kinetic energy when the brakes are applied, improving fuel combustion through lean burn technology and so on.

But perhaps the only realistic long-term option is to reduce the number of cars on the road. That may not seem realistic when forecasts all seem to indicate that there is going to be a big rise in car ownership. The problem is that in some areas now at long last the roads are proving to be inadequate for the number of cars which want to use them, and though it is always possible to build new roads the growing interest in the environment is leading to public pressure against new roads in some countries.

In the United States more than 60 000 square miles of the surface has been paved over – 2 per cent of the total surface area. At least a third of the land in any particular city is devoted to roads and car parks. In Los Angeles two-thirds of the city is devoted to the car. But freedom to own and drive a car is tempered by the inconvenience of getting stuck in traffic jams. In London the average car journey proceeds at eight miles an hour and in Tokyo it is even less. The problem is that building more roads simply encourages more cars and within a decade or so congestion is as bad as ever. Official Government forecasts in Britain show that by the year 2025 the number of car miles travelled in Britain will be up by between 83 and 142 per cent compared with now when the streets are already congested.

In California the average car journey proceeds at 33 miles per hour. By the year 2000 that is expected to drop to 15 miles an hour. Apart from the inconvenience, cars that are stationary in traffic jams continue to burn petrol. It is estimated that in the United States nearly 3 billion gallons of petrol are wasted in this way each year.

The increased congestion is now leading to more demands in many countries for a more effective public transport system and it is no longer just the eccentric fringe who travel by bicycle.

CHAPTER EIGHT

Waste – Disposal on Land

In one way or another man is now producing so much waste that he is in danger of being swamped by it. New York has the world's biggest rubbish dump. It is called the Fresh Kills landfill site and stretches over some 3000 acres. Cranes as tall as six-storey buildings work round the clock emptying barges of waste from the city – 26 000 tonnes of it a day – creating literally a mountain of rubbish. The site will eventually be the highest point of land between Maine and Florida equalling in height some of New York's lower skyscrapers. In less than 10 years' time the site will be full and then the people of New York will have to find somewhere else to dump most of their rubbish. There have even been proposals to bring it to Britain for disposal, though with Britain producing 80 million tonnes of rubbish of her own each year the plan has not been welcomed.

But rubbish, once dumped, does not disappear. The site has to be carefully monitored in case it causes problems – one of the main ones being that as it rots down it produces gas.

In theory, 1 tonne of rubbish can produce 400 cubic metres of landfill gas – 60 per cent of it is methane and around 40 per cent carbon dioxide with a few other trace gases like nitrogen and hydrogen. Typically, landfill gas is produced fairly quickly over the first few years and then production slowly tails off. On average it takes about 15 years for about a quarter of the waste to rot down, so the danger is long lasting.

At most landfill sites the methane and carbon dioxide diffuse into the air. But landfill gas can represent a real hazard. Methane, for example, can explode when concentrations in air reach 5 to 15 per cent.

In Britain there have been more than a dozen cases where homes have had to be evacuated because of the danger of landfill gas seeping into foundations. On 24 March 1986 in Clarke Avenue, Loscoe, Derbyshire, Kathleen Middleton's home was destroyed by an explosion caused by methane gas from a tip 50 yards away. It had seeped into her kitchen and been ignited by her central heating boiler.

Local Authorities have identified some 500 waste tips which threaten public health because of the methane they give off and the Government has come under increasing pressure to tighten controls on these sites. The cost of making them safe has been estimated to be as much as £250 000 per site. The dangers of landfill gas have even led to the setting up of an Emergency Response System by the Harwell Laboratory in Oxfordshire. Experts can be called out day or night to investigate landfill gas problems at commercial and industrial premises which have been built on top of landfill sites.

However, methane produced by rubbish can actually be turned to advantage because it can be collected and burnt to generate power. Exploiting landfill sites to tap their energy potential began only 15 years ago, largely as a response to the oil crisis. America began operating the world's first landfill gas project at Palos Verdes in southern California. Britain followed about five years later. The plans to export some of New York's rubbish to Britain envisaged it being dumped in disused quarries in Cornwall and there were seductive promises that it could be used as a power source – providing methane gas equivalent to 40 megawatts of electricity. But such temptations failed to win the hearts of the residents in the sleepy villages through which the huge lorries carrying the rubbish would have had to go, and who can blame them? The Chairman of the Royal Commission on Environmental Pollution in Britain, Professor Sir Jack Lewis, criticised the idea. In a letter to the Junior Environment Minister at the time, Lord Caithness, he wrote, 'When methane is produced as a by-product of waste treatment or disposal, it may make sense to recover the energy from it, but this does not justify creating such conditions for energy production purposes.'

In 1986 the Department of Energy estimated that of the 500 sites in England and Wales producing methane there were 300 where it would probably be worth collecting the gas given off. It calculated that each year the methane from refuse is equivalent in energy terms to one and a half million tonnes of coal of which at least one million tonnes would be worth exploiting. At the moment fewer than 30 sites are being exploited in this way, though enthusiasm is growing. In 1989, gas equivalent to 250 000 tonnes of coal was burnt, worth at least £12 million.

If waste tips were properly managed it would be possible for more energy to be extracted from them, but that would mean controlling more precisely what goes into them. Non-biodegradable rubbish would have to be excluded and there would have to be controls on how the rubbish was deposited in order to produce the most ideal conditions for a biogas reactor. Sewage sludge could be added to help boost the digestion process. If such techniques were used in conjunction with combustion units the amount of energy derived from rubbish might be increased to the equivalent of 5 million tonnes of coal.

And there's another reason why collecting and burning the gas from landfill sites is worthwhile and environmentally friendly: it reduces the greenhouse effect. Methane is about 27 times more effective as a greenhouse gas than carbon dioxide. That means that if the methane is collected and burnt to form carbon dioxide the net impact on the greenhouse effect is reduced. The 28 million tonnes of biodegradable waste which is thrown away each year produces methane equivalent in its greenhouse effect to 54 million tonnes of carbon dioxide a year. But if most of the methane were to be collected and burnt it would be converted into just 16 million tonnes of carbon dioxide. Burning the landfill gas also breaks down some of the CFCs which will be in the rubbish and so reduces still further the greenhouse effect.

Methane is not the only problem associated with rubbish tips. Another risk is that noxious substances can leak away, poisoning rivers and aquifers. The fact that this has not become a real issue yet in Britain is probably due to the lack of sufficient monitoring facilities round landfill sites. But there have been scandals in other countries. For example, in Spain in the last few years there has been an outcry over the dumping of the residues from the manufacture of the pesticide Lindane in the Pyrenees. Between 1976 and 1989 more than 100 000 tonnes of the chemical hexachlorocyclohexane, or HCH, were dumped in the Pyrenean countryside. According to the World Wide Fund for Nature this has been draining into the drinking water supply and irrigation system covering hundreds of thousands of hectares of agricultural land. Another waste product, dioxin, is thought to have been dumped at sites near the Gallego river system which flows into the Ebro. In 1987 EC experts from Brussels tried to investigate water pollution in the area but the Spanish Government denied the Commission's experts entry into Spain. The Spanish Government did, however, indicate that they would stop the particular company dumping.

Searching for safe places to put waste became something of a *cause célèbre* during the 1980s. On one occasion the freighter *Khian Sea* from Philadelphia cruised much of the Western Hemisphere searching for a place to dump 15 000 tonnes of ash. And there were other even more celebrated wandering waste ships like the *Karin B*, which we shall deal with later.

Dumping waste is not the only option. There are two other possible ways of dealing with it: it can be incinerated or recycled, though currently only a small proportion of waste is dealt with that way.

INCINERATING RUBBISH

Burning rubbish reduces its bulk but can generate poisonous gases and there is still the problem of what to do with the ash. Burning the rubbish releases more of its energy than letting it rot down because some rubbish, like plastic, does not rot. Rubbish has a calorific value of between a quarter and a third of that of coal and can be used as a fuel. Ordinary household rubbish can even be made into fuel pellets. It is a complicated process but there are half a dozen plants in Britain which turn waste into fuel pellets with some degree of success.

One scheme on the Isle of Wight can produce 20 000 tonnes

of pellets a year. The waste is first separated into the lighter combustible material and the heavier fraction by putting it through revolving screening drums. Then after a process of milling and drying it is extruded through 20-mm dies to produce pellets about 50 mm long. In theory, from every kilowatt of energy used in the process you get 16 kW of potential energy from the pellets.

Since the war the nature of household waste has changed so that it is potentially more polluting now than 30 years ago. In the days of coal fires many people burnt much of their waste and ash formed a bigger proportion of the dustbins's load. With the introduction of smokeless zones and the increase in central heating and gas boilers, more household waste finds its way into dustbins. In addition, the composition of the waste has changed. It now contains more packaging and plastics and more biodegradable material. What is more, since the Control of Pollution Act in 1974 the waste disposal authorities have closed down many of the smaller sites to concentrate their resources on fewer larger sites, for economic reasons. The result is that the impact on the environment is far greater if anything goes wrong with them.

RECYCLING

Recycling is another way of dealing with household and other non-toxic waste and now that the green movement is making a bigger impact, recycling is increasing. For example, the amount of waste paper being recycled world wide has gone up from 20 per cent in 1960 to 25 per cent in 1984 (the latest year for which United Nations Environment Programme figures are available). These global figures conceal a big increase in recycling in certain countries which are particularly environmentally conscious like Switzerland and the Netherlands where 40 per cent of waste paper is recycled. In 1984 Britain recycled 29 per cent of its waste paper but that figure is likely to be far higher now following increased awareness of the value of recycling.

The aluminium content of cans is very pure, so recycling

these is particularly worthwhile. In the USA a third of the aluminium which is recycled comes from cans.

Recycling steel conserves energy but usually results in some release of trace metals into the environment. The proportion of steel recycled in relation to that consumed has changed very little over the last decade in many countries.

The use of plastics has increased rapidly. In the USA plastics account for 7 per cent of all solid wastes and may reach 10 per cent over the next decade. It is difficult to sort plastic waste from municipal rubbish but there is considerable scope for increasing plastic recycling, which would save energy and conserve petroleum. And glass recycling is increasing in some European countries. In Britain the glass industry hopes to have 5000 bottle banks in place by the end of 1991 – the ultimate goal being one per 10 000 people.

Interest in recycling is gathering momentum, particularly for household waste. In Britain the Government's aim is to recycle 50 per cent of all recyclable household waste by the year 2000. The segregation of waste by householders will change the composition of the wastes which are eventually sent to landfill.

Ironically, recycling may do little to reduce pollution. In industry, though recycling often leaves behind less waste, it is in a more concentrated form, which causes its own problems.

HAZARDOUS WASTES

It's not just household rubbish which is increasingly having difficulty finding a home. All over the world industry produces cocktails of highly poisonous wastes. As we have seen, in many countries with lax controls some of this is spewed out through chimneys to poison the air, with serious effects on the health of local people. Sometimes it is dumped in rivers and in the sea (see Chapter 9). And sometimes barrels of waste are dumped on some unsuspecting country where they remain as chemical time bombs waiting to release their dangerous load. It was in this way that the saga of the ship the *Karin B* began – an event which drew the problem of hazardous waste disposal to the attention of the world.

KARIN B

In the latter half of 1987, 2000 tons of chemical waste in 10 000 barrels left the Italian port of Livorno for the small port of Koko in Nigeria, where it was dumped in the bush. No doubt the Italian entrepreneur who masterminded the scheme was well pleased with his profit. But the people who lived in Koko did not realise the danger they were being exposed to. And even if they had known, it is doubtful they would have been able to do much to stop it. In June 1988 Friends of the Earth sent a team to Nigeria to examine the waste and found it to be in very poor condition. Six to eight thousand 200-litre drums were stacked on pallets and a further 2000 drums were contained in 30 ship containers. Some drums contained PCBs. Others were rusting and many were ruptured or leaking. The ground was contaminated with oils and resins. The results of the chemical analysis, published in a subsequent report, showed that the drums contained 'some of the most difficult and intractable wastes produced by industry in the developed nations'. The report warned of the danger of a fire sweeping through the waste compound causing explosions among the drums. And it also warned of the possible long-term ill-effects on the health of the people living nearby.

Nigeria promptly asked the Italian Government to take the waste away. It was loaded on to the *Karin B* and then began a long and fruitless search for a final resting place. Country after country turned it away, including Britain. Finally, several months after leaving Nigeria, the Italian Government ordered the *Karin B* back to its port of origin, Livorno, for the waste to be landed. The *Karin B* was only one of several ships carrying Italian industrial waste which Italy finally had to take back when no other country could be found which would accept it. The story of the *Karin B* showed up defects in international controls over the disposal of waste, though the fact that Italy was eventually shamed into taking it back strengthens the views of those who believe that each country should be responsible for its own wastes.

The sort of dumping in Africa which gave rise to the *Karin B* episode is, according to some, just the tip of the iceberg. The head of the United Nations Environment Programme, Dr Mostapha Tolba, has said he knows of a Caribbean country where officials were offered more than $250 million to take 10 000 tonnes of hazardous waste every day for a year. The country in question was only able to incinerate 1000 tonnes a day. What, he asked, would have become of the other 9000 tonnes?

PCB SHIP

Another toxic waste 'scandal', which bore some similarity to the *Karin B*, occurred in August 1989 following a big fire at a Canadian warehouse in the Montreal suburb of St Basil-le-Grand. The warehouse had been used to store PCBs, and after the fire about 100 tons of the PCBs were loaded into the Russian freighter *Nadezhda Obukhova*, the first of several consignments which were to be brought to Britain for destruction. But the plan fell foul of the environmental organisation Greenpeace, which succeeded in whipping up public hostility to the idea of bringing such substances into Britain despite the fact that the 'trade' had been going on for years.

PCBs, or polychlorinated biphenyls, were first produced for widespread use in 1929. There are about 209 different types. They are man-made organic chemicals, ranging from oily fluids to heavy greases and waxes and are very stable. That, and the fact that they do not burn easily, made them very useful as lubricants and insulating fluids. They were mostly used in the production of electrical equipment, particularly transformers. They were also routinely used in carbonless copying paper and newsprint. About one million tons were made until production ceased around 1979. No one knows how much of this total production still exists – one estimate by the OECD in 1987 suggested that there are probably about 300 000 tons still in the United States and 283 000 tons in EC countries.

Concern about PCBs began around 1976 when Swedish research showed that they had a tendency to build up in aquatic animals. It was confirmed that PCBs are indeed very persistent and break down only very slowly in the environment. Over the years large quantities of PCBs have been discharged into the environment with no thought to the damage they might cause. At the time there were no controls on the use of new substances and as a result they are now found everywhere. Their presence can be detected in the atmosphere and in the oceans, in animal and plant tissues and even in human milk. They are concentrated through the food chain and accumulate in body fat. There is no clear indication of exactly what harm they do to humans, but people who have been in closest contact with PCBs – the PCB process workers – have suffered an increased incidence of a type of acne and liver problems.

The cargo of PCBs from the Montreal fire, due for incineration in Britain, was turned back following the fuss that was made of the event in the media. But a number of informed people protested that because PCBs are so dangerous the safest thing to do with them is to destroy them by incineration. If PCBs are incinerated at too low a temperature there is a danger they will produce the poisonous chemical dioxin. Britain is one of the few countries which can incinerate PCBs satisfactorily and in the eyes of many experts it made environmental sense for Britain to destroy them rather than send them away, perhaps to be destroyed in less than ideal circumstances or, worse still, dumped. Because of the outcry the Canadian PCBs were eventually returned to Canada where they were put in storage in Quebec.

The story of the Montreal PCBs differs in one important respect from what happened with the *Karin B*. With the *Karin B* the waste was being dumped with no thought to the consequences. With the PCB shipment the chemicals were to be destroyed in a properly controlled manner. The similarity with the *Karin B* is that in both cases the eventual disposal was being done outside the country of origin. There is an irony in that although Canada was quite happy for its PCBs to be exported to Britain for destruction the country has placed an embargo on the importation of PCBs.

Incinerating hazardous wastes has to be done carefully and at a high enough temperature to ensure that no dangerous combustion products are formed and released through the chimney. Britain has three high temperature incinerators – at the Re-chem International plants at Pontypool and Southampton and at the Cleanaway plant at Ellesmere Port. Following concerns expressed by local people worried about how effective the plants were at destroying wastes, safety tests were carried out in 1984 to see if they burned PCBs safely. The tests demonstrated that the PCBs were destroyed 99.999 per cent efficiently. Levels of dioxins and similar chemicals, known as furans, in the emissions were consistently below the levels of detection.

Concern over the levels of dioxin given off by incineration might just as well be directed to the emissions from ordinary municipal incinerators which burn rubbish at lower temperatures. They have been found to produce low but quantifiable concentrations of dioxins and furans in the emissions from the chimneys. Concentrations between 10 and several thousand parts per billion have been found in tests carried out in various parts of the world. In Britain tests on municipal incinerators have found levels at the lower end of this range. And it is not just municipal incinerators that produce dioxins, similar levels can be emitted by crematoria! But what are dioxins, and why should there be particular concern about them?

DIOXINS

On 10 July 1976 in the town of **Seveso** in Italy a chemical plant exploded and disgorged huge quantities of dioxin – a substance which up till then few had heard of. The explosion had come as a complete surprise because the chemical plant had been shut down for seven hours. The 4½ tonnes of chemicals inside it, which included sodium trichlorophenate and ethylene glycol, were at a temperature of 158°C – well below any temperature at which heat-producing chemical reactions might have been expected to take place. But the chemical engineers had miscalculated. It was discovered a long time later that the vapour above the liquid had been at a higher temperature than expected – high enough to trigger a chemical reaction which itself raised the temperature still further. Eventually the temperature reached 230°C and the plant blew up.

The result was that one of the more dangerous forms of dioxin known as TCDD was spread over seven square miles. It caused a disfiguring form of acne, called chloracne, which affected more than 100 people and helped give dioxins their bad reputation. Dioxins were also a constituent of Agent Orange used in the Vietnam War by the Americans as a defoliant.

Since then there have been many reports in the press of claims that dioxin is responsible for a variety of human and animal conditions. One paper has even claimed that dioxin is 170 000 times more lethal than cyanide.

The truth may well be somewhat less dramatic. A report by British Government scientists in June 1989 looked at all the evidence about the toxicity of dioxins, including the aftermath of the Seveso accident, and concluded that while they are nasty chemicals they are not as dangerous as popular mythology would have it. The report pointed out that dioxins are created by combustion of all sorts of things including household rubbish, coal (domestic coal fires are said to be the second largest source) and by smoking – probably the biggest source of dioxin for smokers. Dioxins are also produced by car exhausts. There was no evidence that the level of dioxins around incinerators was any higher than elsewhere. When the report was published the Government's Chief Medical Officer Sir Donald Acheson was reassuring about the substances. He said: 'While very high doses of dioxin can produce a skin condition known as chloracne and abnormalities of the peripheral nerves, no deaths have ever been reported due to dioxins and there is no good evidence of cancer or birth defects in humans who have come into contact with high exposure to dioxins in the past.'

But the Government report has failed to still all concern. In October 1989 the *New Scientist* carried a report of a conference in Toronto in which a new analysis of data from an industrial accident in a chemicals plant in 1953 seemed to indicate that dioxins might have been responsible for an increased incidence of cancer among those exposed. As with so many potentially toxic substances there is a lack of knowledge on the long-term effects on mankind of continuous exposure to exceptionally low doses. The effects, if any, of such exposure could well take many years to discover and only then by the most careful studies of the patterns of disease in large enough populations.

The principle that each country should be responsible for its own wastes should take into account the possibility that some countries may be better equipped than others at disposing of them. With all countries sharing one environment it surely makes sense that international agreements should allow for the trade in certain wastes if the end result is to be a cleaner and safer world for all. As far as Europe is concerned there is now an agreement that every member of the European Community should be self-sufficient in dealing with waste.

A LESSON FROM THE RECENT PAST

Perhaps it is more widely appreciated now than ever before that the disposal of hazardous wastes by incineration is probably the most effective way of getting rid of them. One of the first manifestations of the problems that had been building up over the dumping of hazardous waste came in the 1970s in a place called **Love Canal** in the town of Niagara Falls in New York State. After one heavy rainstorm residents noticed large fuming pools of chemicals in their gardens. There had been other reports of rusting drums surfacing in gardens, and on one occasion some of the turf at the local school baseball pitch disappeared into a drum which had been just beneath the surface. The legacy of improper waste disposal thirty years before was finally being recognised. A chemical company had dumped many drums of chemicals into an old canal in the early fifties and buried them under earth. But years of heavy rains had caused the canal to overflow and lift the barrels to the surface. Eighty-two chemicals were eventually identified, 11 of which were thought to be capable of causing cancer.

The area was evacuated, but for some it was too late. There was evidence that already the area had a poor health record with higher than average levels of cancer, liver disease and respiratory ailments, birth defects and retarded children. The clean-up operation was extensive and expensive. Now a drain delivers water from the contaminated area to filters. In 1972 legislation was passed which regulated the disposal of dangerous wastes. But the story of Love Canal may well be repeated in many other areas until waste tips are cleaned up.

If America has its problems, Britain has not exactly set an example to the rest of the world in the proper control of waste disposal. Much of the current practice dates from the 1974 Control of Pollution Act which was designed to regulate the disposal of what are called 'controlled wastes'. It led to the setting up of a number of Waste Disposal Authorities across the country to operate waste disposal sites and license private sites. Since then a number of reports from, among others, Parliamentary Select Committees and from the Royal Commission on Environmental Pollution have been strongly critical of Britain's waste control policy.

Under attack in particular have been the practical arrangements of waste disposal described as 'ramshackle and antediluvian', and attention has been drawn especially to the practice of 'co-disposal landfill' – the burying of hazardous wastes along with domestic refuse – a potentially dangerous policy, and one which is not legally permissible in the United States.

Until the end of 1988 there were still only six inspectors to oversee more than 5000 disposal sites in England and Wales. This left the system open to abuse. As an example, one case which came to light involved a contract which was signed at £1350 for the incineration of some dangerous research waste which was stored in nitrogen in special pressure vessels. The waste was particularly dangerous as it was spontaneously flammable in contact with air or water. The waste disposal contractor subcontracted the disposal to the operator of a land based waste store who then further subcontracted to the operator of a waste tip at a price of just £75. The chain was discovered and the authorities stepped in. Had the waste been dumped the results could have been disastrous.

Britain's waste disposal policy suffered yet another stinging criticism in February 1989 in the second report of the House of Commons Environment Committee. While the 1974 Control of Pollution Act referred to the disposal of what it called 'controlled wastes', the Environment Committee's report made it clear 15 years later those wastes had still not been properly defined. As for the Waste Disposal Authorities, the Commons Environment Committee's report said too many took their dual role of poacher/gamekeeper too lightly, 'appointing insufficiently qualified officials low in the establishment pecking order and without the clout to demand an adequate allocation of resources for their highly technical responsibilities'. It was pointed out that as far back as 1979 legislation had required local authorities to submit their plans for the disposal of controlled wastes in their areas to the Department of the Environment. Yet ten years later 56 out of the 79 English Waste Disposal Authorities had still failed to do so. The Department of the Environment meanwhile seemed to have done nothing about it.

Among the other criticisms in the report was one aimed at the chemical industry which, the MPs said, lacked commitment to research on wastes. It pointed out that in the previous ten years no new technologies had emerged to deal with wastes. While the best waste disposal companies operated to a high standard the industry was still immature, said the report, and the worst sites 'are appalling and potential disaster areas'.

In its reply to the Environment Committee's report the British Government acknowledged some problems. But it also claimed the Committee had gone too far in some of its criticisms and that its concerns about the Control of Pollution Act had not been supported by evidence. In particular the Government claimed that the Committee had found little evidence of pollution from landfills.

The Government said it did not accept that Britain's waste management standards posed a serious risk to the public or to the environment. Improvements to waste management had already been put in hand when the Committee began its deliberations. As for co-disposal the Government maintained that co-disposal for a range of wastes was technically sound. The Government made it clear that they intended to introduce controls over those who produce or control wastes. In future they would have what is called a 'duty of care'. In other words they would be required to take reasonable care to ensure its lawful disposal (all except private householders). That duty of care was finally enshrined in the Environmental Bill of 1989. It means, among other things, that waste management licences will remain in force even when the waste disposal facility is full. Whoever holds the licences will remain responsible for making sure the waste does not cause any problems perhaps for up to 50 years into the future. Looking after waste will become an ongoing responsibility.

The Environmental Bill also contains proposals for what is called Integrated Pollution Control. Waste disposal from the 3300 operating processes which produce potentially polluting waste will have to be disposed of using what is known as the 'best available techniques not entailing excessive costs' or BATNEEC. And those techniques are to be regulated to minimise the effects on air, water and land.

Another aspect of the Bill which has been widely welcomed is the provision that 'special' wastes will be defined by reference to eleven characteristic properties which make the waste difficult or dangerous to dispose of. That will include substances which represent a threat to the environment as well as to human health.

INTERNATIONAL CONTROLS ON SHIPMENT OF HAZARDOUS WASTES

It is inevitable that there should be trade in international wastes so long as there are some countries which cannot deal with them effectively on their own. Britain, with three high technology incinerators, has increased its imports of special wastes from 4000 tonnes in 1981 to 52 000 tonnes in 1986–7. While that may sound a lot, and has stimulated environmentalists to call for a halt, it is still only 4.5 per cent of the special waste which Britain produces herself. France, by comparison, imports about 200 000 tonnes a year. The OECD has estimated that as long as five years ago the amount of hazardous waste which was crossing national frontiers in Europe alone was more than 2 million tonnes – much of it from west to east. Indeed, the Federal Republic of Germany recently disclosed that it alone shipped 600 000 tons of hazardous waste each year to East Germany.

For a long time there were no controls over the international shipments of waste. But following the Seveso disaster in Italy in 1976, 41 drums of dioxin were temporarily lost between Italy and France and as a result of this, in December 1984, the EC adopted the first Transfrontier Shipment of Hazardous Wastes Directive. Eighteen months later in June 1986 the directive was amended to provide more precise rules governing the export of hazardous waste from Member States to third states. Included in the definition of hazardous wastes are PCBs and certain other toxic and dangerous wastes which exist in sufficient quantities as to pose a threat to health or the environment.

The directive stipulates that before waste is transferred, the authority which is going to dispose of it must be told what the waste is and what its characteristics are. It must also be told which facility is going to dispose of it. The disposal authority has the power to stop the shipment if it is dissatisfied with the arrangements. The directive also stipulates that the wastes must be covered by full documentation from beginning to end.

Other controls on the international movement of wastes are being worked out within the OECD and also within the United Nations Environment Programme.

In America concern about hazardous wastes has led the Government to insist that all hazardous wastes should not be dumped but eliminated. It is estimated that American industry generates more than a ton of solid hazardous waste per head of population each year. About 10 000 sites containing hazardous wastes will need to be cleaned up – a job which will cost probably more than $100 billion.

The decree has led to a boom in the waste disposal business as various technologies vie with each other to prove they are the most effective way of destroying the wastes.

CHEMICAL TREATMENT OF WASTES

Burning waste is not the only way to get rid of it. It can also be treated chemically to make it less harmful. At present this is not often done, partly because it is so expensive, but at Dounreay in Scotland scientists have developed an electrochemical process which they claim has considerable potential for destroying a wide variety of organic wastes including PCBs and chlorinated solvents. The process involves passing an electric current through a solution containing the chemicals to be destroyed. It operates at a relatively low temperature reducing the possibility of anything going wrong which might cause the waste to evaporate and spread into the environment. It can be switched off in seconds and it can be used with different types of chemical waste which, the scientists say, removes most of the problems associated with poor characterisation of the chemicals to be destroyed. Because it consumes power it is more expensive than incineration but it could prove valuable for some of the more dangerous wastes.

In this chapter we have been mainly concerned with the problems of what happens to waste that is disposed of on land. But a large amount of chemical and other waste is disposed of by piping it into rivers or dumping it at sea, as we will see in the following chapter.

CHAPTER NINE

Waste – Disposal in Water

Water is one of the most basic necessities of life. We drink it, wash in it and cook in it. Animals and plants cannot live without it. But increasingly over the last hundred years water has become prone to pollution as the effluent from towns and cities, from industries and from agriculture, is discharged to rivers, seas and lakes and contamination seeps down to underground aquifers. Over the years pollution has killed animals and plants, and has made many rivers biologically dead. River-born pollution has brought with it disease and death for man.

In the last twenty years or so man has woken up to the damage being done and, slowly, national and international controls over what can be allowed into rivers, seas and lakes are being implemented. Cleaning up the world's rivers and seas will be a painfully long and expensive business but all those who have studied the problem agree that it is essential.

So how sick are the world's seas? And what is being done to protect them?

From all parts of the globe there are reports of polluted beaches and coastal waters. Coastal fisheries are threatened as is the tourist industry in many countries. To take just a few examples: in the summer of 1987 hundreds of dead dolphins were washed up on the coast of New Jersey, along with raw sewage and other rubbish. Their deaths may have been due to natural causes but the suspicion is that pollution played a significant role. In the same year a third of the United States' shellfish beds were closed because of the risk of pollution. The annual harvest from Long Island's clam and scallop beds, worth $110 million declined to less than half that amount. In

Louisiana since 1982, 70 per cent of the state's oyster beds have been closed for up to six months of the year because of pollution by sewage. In 1973, 14.7 million lb of striped bass were caught in New York's waters. In 1986 bass fishing was banned because the fish were found to be contaminated with polychlorinated biphenyls.

The sea has been used as a dumping ground for centuries. The oceans have swallowed up millions of tons of waste a year and, in the past, few people gave much thought as to the effect the waste might have on the oceans. The oceans are so vast that there was a tendency to regard them as a bottomless cess-pit capable of absorbing anything man could put in them, including the sludge resulting from sewage treatment, the tailings left over from mining operations, wastes from the chemical industry, and ash from power stations.

Twenty years ago it became clear that some sort of assessment had to be carried out and controls enforced to slow down and if necessary stop the dumping process. The first major international meeting to examine the problem took place in London in June 1971 as part of the preparations for the first United Nations Conference on the Human Environment in Stockholm the following year.

An Inter-Government working group on marine pollution was set up and in the end the Stockholm meeting agreed that governments should ensure that: 'ocean dumping by their nationals anywhere, or by any person in areas under their jurisdiction, is controlled and the governments continue to work towards the completion of and bringing into force as soon as possible of an over-all instrument for the control of ocean dumping . . .'

As a result of the Stockholm meeting Britain convened a conference which met in London later in 1972 and adopted what is called the 'Convention on the Prevention of Marine Pollution by Dumping of Wastes and Other Matter'. The convention, which has become known as the **London Dumping Convention**, entered into force in 1975. By 26 July 1983, 63 governments had ratified or acceded to the convention. The effect of the London Dumping Convention has been to make clear the exact quantity and type of waste dumped at sea, and what should be done to reduce it.

Waste is divided into three groups. The most environmentally dangerous materials are listed in an annex to the convention and their dumping at sea is banned. They include organochlorine compounds, mercury, cadmium, persistent plastics, crude oil, high-level radioactive wastes and materials produced for biological or chemical warfare.

There is a second list of substances which may be dumped only after a prior special permit has been granted. Included in this list are arsenic, lead, copper, zinc, organosilicon compounds, cyanides, fluorides and pesticides.

Other material, such as dredgings, may only be dumped after what is called a prior general permit has been granted. There are other provisions in the agreement governing such things as how the material is packaged and where the dumping is to take place.

Two other international agreements which help control pollution of the sea arose out of an incident in 1971 when the coaster, the *Stella Maris*, left Rotterdam with a cargo of chlorinated hydrocarbons for dumping at sea. It was forced by public opinion to move from one site to another. This and other similar incidents brought European countries together to tackle the general problem in their area and their deliberations eventually gave rise to two conventions: the **Paris Convention** (1974) which regulates discharges from land, and the **Oslo Convention**, applying to the Arctic and northeast Atlantic, which regulates the Marine Pollution by Dumping from Ships and Aircraft (1972). These conventions are more specific than the London Dumping Convention which is world wide. In addition there is the Bonn Convention which deals with oil pollution.

The three main materials dumped at sea are dredgings, industrial waste and sewage sludge.

1. DREDGINGS

Ninety per cent of what is dumped at sea is material which has been dredged from the sea bed. Two-thirds of it has come from harbours, rivers and other waterways to keep them from silting up and the other third is as a result of new works. The developed world dumps on average 215 million tonnes of dredged material a year at sea though the figure is declining. It might sound reasonable and fairly harmless to dump dredged material at sea but in fact about 10 per cent of it is heavily contaminated with waste from shipping, with industrial and municipal waste and with land run-off. The typical contaminants include heavy metals like cadmium, mercury and chromium, hydrocarbons like oil, and pesticides and nutrients including nitrogen and phosphorus. There is a risk that if these are disposed of over the open sea they could damage marine organisms and get into the food chain. Even dumping uncontaminated dredged material at sea is not without its hazards to the environment because it may well drift down to cover the gravel bed habitats of crustacea and of spawning fish like herring. So selecting a suitable site for such dumping is of key importance.

2. INDUSTRIAL WASTE

Several million tons of industrial wastes are dumped at sea each year. Most of it is acid and alkaline waste, scrap metal waste, fish processing waste, coal ash and flue gas desulphurisation sludges. There has been a steady decline in the dumping of industrial waste. In 1982 the developed world dumped 17 million tonnes of this waste into the sea but by 1985 that figure had been reduced to some 6 million tonnes.

3. SEWAGE SLUDGE

Sewage sludge can be useful as a fertiliser on land but much of it is still dumped at sea, especially in the southern North Sea, parts of the Irish Sea and in the New York Bight. Though sewage sludge does not usually contain large amounts of contaminants it does contain a lot of nutrients which can cause what is called eutrophication – the rapid and uncontrolled growth of plant life – and it may also contain disease-causing organisms. Between 1980 and 1985 developed countries were dumping an average of 15 million tonnes of sewage sludge at sea each year though, like industrial waste, that figure too is declining. The International Maritime Organization believes that sea dumping of sewage waste may still be the best environmental option for those countries without advanced treatment methods.

As far as Britain is concerned 30 per cent of sewage sludge is dumped at sea – some 5 million tonnes a year. Of the rest, about half if spread on agricultural land or used as a soil conditioner for land reclamation, another 16 per cent is transported to landfill sites, sometimes mixed with municipal refuse, and 4 per cent is incinerated. The Royal Commission on Environmental Pollution in 1985 considered that disposing of sewage through outfalls into shallow water or by direct dumping was the best practicable environmental option. But the practice has been increasingly questioned by environmentalists.

Now a major clean-up of sewage is on the cards in Europe following a draft directive from the European Commission which lays down much tougher standards than Britain has generally adopted. The Directive means that where sewage outfalls serve a population of 10 000 or more the sewage will normally receive what's called secondary treatment*.

The Directive makes allowances for some sewage to receive only primary treatment where the water into which the sewage is being pumped can tolerate it and occasionally tertiary treatment may be necessary to meet specific environmental needs.

In Great Britain and Northern Ireland there are 151 sewage outfalls discharging to estuaries or tidal rivers and 102 outfalls discharging to the sea. Quite a lot of the sewage which goes through them may have had only preliminary treatment or even none at all. The British Government announced in March 1990 it was accepting the draft Directive. The Government

* (Preliminary treatment effectively sieves the sewage and removes solids like grit and plastic, primary treatment involves holding the sewage in sedimentation tanks allowing a sludge to settle out, secondary treatment allows biological oxidation of the effluent from primary treatment and tertiary treatment improves the effluent from secondary treatment using such things as sand filtration and nutrient removal.)

accepted that the policy of relying on long sea-outfalls to disperse sewage had never been enthusiastically accepted by the public. In addition not all scientists agreed it was the best way of dealing with the problem and there had been concern about the long-term build up of pollutants in the sea. As far as bacteria and viruses are concerned, in Britain at least, there have been no major outbreaks of serious disease as a result of polluted sea water but it can lead to swimmers developing minor infections of the ear, nose and throat and certain skin infections. And aesthetically the sight of sewage slicks, even if they are well out at sea, does not encourage confidence in the resort!

INCINERATION AT SEA

Another possible cause of marine pollution comes from incinerating hazardous liquid wastes at sea. The practice began in 1969. Between 1980 and 1985 an average of 100 000 tonnes of hazardous waste were incinerated each year, mainly in the North Sea. In 1985 the waste incinerated in the North Sea came from the Federal Republic of Germany (58 000 tonnes), non-North Sea countries (17 000 tonnes), Belgium (13 000 tonnes), France (10 000 tonnes), the Netherlands and Norway (3000 tonnes each) and Britain (2000 tonnes).

One direct result of incineration at sea has been that significantly increased levels of a substance called HCB (hexachlorobenzene) have been measured in the water near areas where incineration has taken place. (It has also been found in even greater concentrations near coasts, caused by river input and dumping.) HCB is one of the chlorinated hydrocarbons – a group of synthetic chemicals which degrade very slowly and which can accumulate in fatty tissues of animals and humans. Many of the substances are toxic and can cause cancer.

An addendum to the London Dumping Convention was adopted in 1979 to control incineration at sea. The regulations make it clear that it should not be seen as an alternative to finding other, better ways of disposing of waste. A group of experts from the London Dumping Convention and from the Oslo

Commission met in April 1987 to examine the safety and environmental acceptability of this practice and they urged further research, especially into such things as the efficiency of marine incinerators and the effect on the marine ecosystems. There is now general agreement that incineration at sea should be phased out by 1994, assuming that other better methods of waste disposal can be found.

So in view of all the pollutants which have insulted the seas for hundreds of years how have they fared? Are they on the verge of becoming lifeless or have they been able to absorb the pollution while still maintaining vigorous health? According to most experts the seas have coped remarkably well, though in a few places there are signs that the ecosystem is suffering.

THE MEDITERRANEAN AND ADRIATIC

More than 100 million people live in coastal regions of the Mediterranean. It is the world's most popular tourist area with 30 000 miles of shoreline. Within 40 years it is expected that nine-tenths of the coast will be developed for tourists. The Mediterranean is virtually enclosed and it has little tidal flow, the waters of the sea are replaced only every 100 years or so. It is bordered along its north coast by some of the world's most industrialised countries and to the south by countries which will one day become industrialised. In general, in terms of nutrient content, the Mediterranean is a fairly poor sea which is why it is so crystal clear and blue. But what it lacks in naturally occurring nutrients it more than makes up for in pollutants. It receives an astonishing amount of mankind's detritus each year – 430 billion tonnes of it including sewage, domestic waste, detergents and chemicals.

Each year coastal industries alone discharge into the Mediterranean 5000 tonnes of zinc, 1400 tonnes of lead, 950 tonnes of chromium, and 10 tonnes of mercury. Probably a further 100 tonnes of mercury are added to this total by rivers, and up to 400 tonnes of mercury are thought to be deposited into the sea from the atmosphere. (Some fish in the Mediterranean contain three times as much mercury as fish caught in the Atlantic. But contaminants like this cannot always be put down to pollution. Ironically the mercury seems to come mainly from natural sources because museum specimens of fish caught nearly a century ago show similar levels.)

The atmosphere is probably responsible for the biggest input of lead. And then there are the organic pollutants like PCBs, persistent pesticides (90 tonnes a year) and oil. A quarter of the world's tanker-borne oil, over 400 million tonnes a year, crosses the Mediterranean and it is estimated that up to a million tonnes of oil are discharged into the sea each year (see Table 5).

TABLE 5 ACCIDENTAL OIL SPILLS FROM TANKERS: 1975–1985

Year	*Number of oil spills*	*Tonnes of oil spilled*
1975	45	188 042
1976	29	204 235
1977	49	213 080
1978	35	260 488
1979	65	723 533
1980	32	135 635
1981	33	45 285
1982	9	1 716
1983	17	387 773
1984	15	24 184

Source: *UNEP Environmental Data Report*, 1989/90 (Causes of spillages including grounding, collision, explosions, heavy sea, missile attack.) ©

The Mediterranean also has an astonishing amount of sewage pumped into it from the countries round its coasts – 90 per cent of which is untreated. Pollution of the Italian coastline is particularly bad. A survey in 1987 found the Bay of Naples so fouled with raw sewage that one of Rome's marine scientists described it as 'on the outer limits of the imaginable'. More than a quarter of the beaches surveyed between 1975 and 1981 had to be closed for a time because of pollution.

Italy's largest river, the Po, is a major source of contamination. It flows for 680 kilometres covering about 25 per cent of Italian territory, 36 per cent of Italy's population and 200 000 industries. About 80 per cent of the organic pollution in the northern Adriatic Sea can be traced to the Po. In the Adriatic eutrophication is the most critical problem, with excessive quantities of nitrogen and phosphorus leading to huge growths of marine algae which dies and rots and ruins miles of beaches.

In 1989 the Adriatic coast was hit by a particularly large algal bloom which covered the beaches with a thick brown gelatinous mess and made swimming in the sea impossible. For years the Po had been discharging nitrates and phosphates from farmland into the Adriatic. Now the people of the seaside resorts were reaping the sorry rewards for the years of neglect by their fellow countrymen. For an area which caters for 40 million tourists a year the impact was enormous. The 1989 tourist industry in the immediate area was ruined. The Adriatic accounts for 20 per cent of Italian fish catches so if the eutrophication persists fishing, too, will be in jeopardy.

Sewage, apart from containing nutrients, may well contain other more serious contaminants such as disease-causing organisms. They can concentrate in sea-food making it unwise to eat raw sea-food around the Mediterranean.

In March 1989 a report from the United Nations Environment Programme, produced by a team of 21 scientists, expressed concern over links between faecal contamination of sea water in coastal regions and disease among bathers – the relationship is particularly strong in the case of children under five. The scientists said the danger is greater in warmer climates, such as in the Mediterranean, where people may spend much longer in the water.

The United Nations has led the way in trying to clean up the Mediterranean. In 1975 the UN Environment Programme launched a number of projects, the two main ones being the Mediterranean Action Plan and the Blue Plan for the Mediterranean. The Action Plan was drawn up following the Barcelona Convention which was signed by 18 Mediterranean countries. The Convention was accompanied by four protocols including one on how pollution should be tackled. Negotiating the protocol was not easy because there were competing interests. The countries in the north wanted to protect their beaches and their tourist industries but were concerned about the expense of fitting existing factories with technology to clean up the effluents. In the south there were countries which did not want to slow up the industrialisation process by imposing extra expense on their industries. In the end a compromise was reached. It was decided to base controls on water quality standards. That meant that countries in the south with relatively clean waters were allowed to industrialise without worrying too much about their effluent while countries in the north, which had already polluted their coastline, bore greater costs. In addition, to ease the problem for northern countries faced with huge bills, it was decided that, with the exception of those factories which were particularly bad polluters, clean-up technology would have to be fitted only to new factories. These compromises persuaded all countries (except Albania) to sign.

Two of the priority areas were identified as the contamination of shellfish by human effluent and the high mercury levels. A meeting in Athens in 1987 agreed to limit discharges of mercury to the sea and also agreed on ways to cut the discharge of raw sewage. In September 1989 the World Bank, in one of its first big investments in the region, paid $218 million towards expanding the sewerage facilities of Istanbul. This will go a long way towards reducing pollution in the Eastern Mediterranean. Other big sewerage treatment schemes are underway in Alexandria, Athens, Genoa, Marseilles, Naples, Tel Aviv and Toulon. Mediterranean countries have agreed that within ten years adequate sewerage systems will be built for all communities with more than 10 000 inhabitants.

THE NORTH SEA

Unlike the waters of the Mediterranean the North Sea is not a closed sea. It is 'washed' by the waters of the Atlantic and the water in the North Sea changes roughly every two or three years. Yet even here there have been signs that pollution carried into the sea by Europe's rivers may be slowly upsetting the sea's ecosystem.

In 1984 the first Conference on the Protection of the North

Sea was held in Germany and a report was produced on the health of the North Sea, known as the North Sea Quality Status Report. It found that more than 70 million tonnes of sewage sludge, industrial waste and harbour dredgings were dumped in the North Sea each year. Rivers brought down at least 50 000 different chemicals from factories and the land, many of which were only dimly understood. They included heavy metals, PCBs, stable chlorine-containing chemicals used in wood preservatives and pesticides. In addition, large quantities of fertilisers – 1.5 million tonnes of nitrates and 100 000 tonnes of phosphates – went into the sea each year. Britain contributed 20 per cent of pollution brought down by rivers and 15 to 30 per cent of the input of dredged materials. Britain was also the main source of contaminants from direct discharge and the only source of sewage sludge. The Belgians and the West Germans were the main contributors to pollution by incineration at sea.

Despite all the pollution the Quality Status report concluded that on the whole the North Sea is standing up rather well to the contamination. The report did identify localised problems, though, notably in the shallow Eastern waters of the German Bight, the Skaggerak and the Kattegat.

Some scientists hold the view that because there is no proof that the North Sea is dying it must be all right. But, as with other aspects of environmental degradation like the greenhouse effect, the hole in the ozone layer and the effect of acid rain on lakes and forests, waiting until there is definite proof could mean leaving things until it is too late. Few would argue with the general idea that preventing pollution must be better than having to clear it up once it has occurred.

There have been many research projects which have sought to identify the effects of pollution in the North Sea. One comprehensive project carried out in 1986–7 by West German researchers found, among other things, a conspicuously high rate of disease in fish around the Dogger Bank – rates previously only seen in polluted coastal waters. There have also been important changes in the range of species found on the bottom of the sea. Thirty years ago there were diverse species on the Dogger Bank. More recent surveys have found only a few worm species which are more tolerant of unfavourable environmental conditions.

Investigations have shown relatively high lead and cadmium concentrations in the central North Sea. And the pesticide, Lindane, which is still being used in several countries bordering the North Sea can be detected in the entire North Sea at levels three to four times higher than in the Atlantic. In the Skagerrak it has penetrated 50 cm down into sediments two hundred years old.

In some ways things are better than they were in the past. The Quality Status report showed that compared with twenty years ago Britain's rivers are much cleaner now and fewer chemicals are being dumped into the North Sea. But there is still a long way to go. The Thames, for example, though much cleaner than it used to be still carries waste from some 136 different industrial pipelines, and contaminated sewage from some 4170 pipelines. Thames Water dumps 4.5 million tonnes of treated sewage sludge into the Thames Estuary annually, which means among other things that 1.5 tonnes of mercury goes into the estuary each year.

The second North Sea Conference was called in 1987 and used the North Sea Quality Status Report as its key document. It resulted in a number of agreements including one to cut by 50 per cent the input to rivers of dangerous substances by 1995, to reduce ocean incineration by 65 per cent and phase it out altogether by 1994 and to end all ocean dumping of 'toxic' waste by 1 January 1989 (though dumping of inert substances would still be allowed).

Although pollution is being reduced some environmentalists claim that the problem is not being tackled quickly enough. In 1988 there were three events, each of which might be interpreted as a symptom of what could be a far greater malaise:

1 In May there was a devastating algal bloom in the Skaggerak and Kattegat – one of the areas identified as at risk in the Quality Status report. Algal blooms can occur naturally if the conditions are favourable but it is thought that this bloom

was the result of a combination of very warm weather and chemical fertilisers washed there by North Europe's rivers. It formed a slick up to 100 feet deep and 6 miles wide. Currents carried it along the Swedish and Norwegian coasts and as it went it killed millions of fish. Before the algal bloom died out it had caused £120 million worth of damage to the Norwegian fishing industry alone.

2 Then there was the much publicised death of seals, first on the shores of Denmark and then off the Wash. At first environmentalists claimed the deaths were caused by pollution but later scientists found the illness was caused by a virus. Whether the seals were made more vulnerable to the virus by chemical pollution is not known.

3 Finally, the population of puffins, arctic terns, arctic skuas and red-throated divers on the Shetland Islands were found to be at a very low level. The reason was identified as a different kind of phenomenon, not pollution this time but an equally important environmental impact – over-fishing of the sand eel, one of the birds' main foods.

These three instances of imbalance may be relatively insignificant when considering the overall health of the North Sea. Some or all may even have been due to natural causes because there are natural variations in the balance of nature. But many people believe they were more likely the result of man's interference with nature and demonstrate that unexpected local disasters can follow careless exploitation of resources.

Clearly it is one thing to control dumping at sea and quite another to control what is carried into the sea by rivers. British rivers contribute 20 per cent of North Sea river-borne pollution. But most comes from other North European rivers – especially the Rhine and the Meuse.

Environmental organisations, especially Friends of the Earth and Greenpeace, have consistently campaigned for greater commitment by Britain to cleaning up discharges into rivers. Though targets have been set, it is still too easy, they claim, for polluters to gain consent to discharge dangerous chemicals into water. And when the limits are exceeded legal action is seldom implemented.

There is a difference of opinion between Britain and some of its European partners on the subject of how rivers can best be cleaned up. Britain's Water Act passed in 1989 introduced the concept of statutory quality objectives which now form the basis of Britain's own system of water pollution control. The continental concept is to set down uniform emission standards for industry. Britain believes that will not ensure the clean-up of the major European rivers like the Rhine, the Elbe, the Weser and the Scheldt which discharge sewage, nutrients and detergents into shallow waters where they do a lot of damage.

It might have been expected that, following the agreements reached after the second North Sea Conference, Britain might well have ended sea dumping of controversial material. As in other ecological matters, however, progress has been slow. The Government announced in February 1990 that the dumping of liquid industrial waste and fly ash into the North Sea would be stopped by 1992. But other European countries have been highly critical of Britain for failing to implement the undertaking given in 1987 that dumping would cease by 1989 – especially given that other countries have stopped.

Following public pressure, and anticipating criticism at the third North Sea Conference in March 1990, the Government also announced that it was to stop the dumping of sewage sludge into the North Sea by the end of 1998. It is now investigating alternative methods of dealing with it. And it announced that raw sewage, currently being discharged to sea via long outfall pipes, would eventually all be treated. That would involve spending £1.5 billion for treatment works.

Despite the expected criticism of Britain, the 1990 conference did reach agreement on a number of ways to cut pollution further. On nutrients, for example, the conference accepted Britain's view that because some parts of the North Sea are more at risk than others, it would not be reasonable to insist on the same strict controls for everyone, regardless of the state of the sea. Only in vulnerable areas will nutrient inputs from sewage, for example, have to be cut by half, involving expensive secondary treatment.

Another issue discussed was the fate of PCBs. Britain's view prevailed: that they should be destroyed by high temperature incineration rather than stored as the Germans wanted.

The Mediterranean and the North Sea are by no means the only seas being adversely affected by dumping and by discharge from rivers and outfalls. One of the most serious problems is occurring in Soviet Russia. According to Professor V. Loukjanenko of the USSR Academy of Sciences, 40 cubic kilometres of waste a year are being discharged into the waterways without any treatment. And the relatively low level of waste-water treatment generally means that only 90 per cent of organic substances and between 10 and 40 per cent of inorganic compounds are removed. This results in significant amounts of nitrogen, phosphorus, potassium and nearly all mineral salts, including the salts of toxic heavy metals 'slipping through' waste-water processing. Writing in the magazine *World Health*, published by the World Health Organisation, he gave as an example what is happening in the sea of Azov. Between 1983 and 1987 the annual concentration of pesticides increased five times. Figures for 1988 show that the average concentration of stable organochlorine pesticides in suspension in the sea has risen 17 times and by a factor of 27 in the Gulf of Taganrog. In the country's important fish breeding areas, the Volga-Caspian basin, 40 per cent of the country's fish and 90 per cent of the world's sturgeon are caught. Each year 367 000 tons of organic waste, 13 000 tons of oily waste, 45 000 tons of nitrogen and 20 000 tons of phosphorus are discharged into the Volga. High levels of organochlorine pesticide residues have been found in sturgeon and sprat. He writes, 'We have to recognise that the path we have pursued for the past 30 years in our quest for "cheap" ways of dealing with water pollution is at a dead end, for the original assumption that water would cleanse itself was fundamentally mistaken and devoid of any scientific basis.' He adds, 'the overall state of the Soviet Union's seas, rivers, lakes and reservoirs can only be described as extremely serious.'

RADIOACTIVE WASTE

Low level radioactive wastes have been dumped at sea since 1946. Between 1946 and 1967 the United States dumped about 4000 units (terabecquerels) of radioactive waste in 90 000 containers of various types in the Pacific and Atlantic Oceans and the Gulf of Mexico. And between 1949 and 1982 several West European countries, mainly the United Kingdom but also Belgium, the Netherlands and Switzerland, dumped about 54 000 units of radioactivity in some 140 000 tonnes of packaged waste at ten sites in the northeast Atlantic. Japan carried out a small amount of dumping between 1955 and 1968 and Korea between 1968 and 1972. The waste came mainly from activities associated with the generation of electricity from nuclear power stations and also from the use of radioisotopes in industry and medicine. It was packed in concrete-filled drums and consisted mainly of items like old clothing and bits of machinery contaminated with radioactivity. The total amount of radioactivity added to the oceans – 60 000 units – is much less than the 200 million units added as a result of the atmospheric testing of nuclear weapons between 1954 and 1962. Even this amount pales into insignificance compared with the amount of radioactivity present naturally in the oceans – estimated at some 20 billion units. But because the radioactive isotopes are different, the total measure of the radioactivity is only a very rough guide to the risk the isotopes represent.

According to the London Dumping Convention the dumping of high level waste is completely banned, but the dumping of low level waste is permitted under certain conditions.

In 1983 a moratorium on dumping low level radioactive waste was adopted pending a review. This review concluded that the risk to individuals from previous ocean dumping is extremely small. The risk of developing a fatal cancer or severe hereditary defect as a result of previous dumping was put at one chance in a billion per year. On the other hand, the experts calculated that the total casualties from past dumping could reach 1000 over the next 10 000 years or so. This would mainly come from radioactive carbon 14 which would over time escape into the atmosphere as carbon dioxide.

As a result of this report it was agreed to ask contracting parties to the London Dumping Convention to suspend radioactive dumping pending more research. As far as Britain is concerned it is unlikely that any further dumping of nuclear waste will occur as the Government has decided that all nuclear waste is to be buried deep underground. Meanwhile until such a repository is built the low level waste builds up at places like Harwell where huge warehouses are stacked high with barrels.

Recently various countries have shown interest in the possibility of disposing of high level radioactive waste into the sea bed. The London Dumping Convention says nothing about this because the technique was not feasible in 1972 when the convention was drawn up, but in 1986 a consultative meeting of the Convention agreed that this should not take place until it is proved to be technically and environmentally sound. At the third North Sea Conference most European countries wanted to stop the possibility of under-sea dumping ever being considered, but Britain insisted that the option for investigating this method of disposal should remain open.

RUBBISH

There is another type of waste dumping at sea which causes considerable harm to wildlife – the dumping of litter and other types of rubbish from ships and offshore installations. It is a form of waste which all holidaymakers come up against when it is washed up on beaches. Under the International Convention for the Prevention of Pollution from Ships this type of dumping is now placed under tight control. The dumping of plastics including **synthetic fishing nets** has been banned, and the dumping of much other waste is banned within sight of land.

The amount of rubbish dumped into the sea from ships and offshore activities is quite astonishing. The US National Academy of Sciences has calculated that in 1975 ships dumped 5.6 million tonnes of rubbish over the side. In 1982 another estimate suggested that merchant shipping disposed of 639 000 plastic containers a year, together with 426 000 glass containers and 7 million metal containers. Of particular concern are fishing nets, especially drift nets, which are made of synthetic material and which can be up to 27 kilometres long. It has been estimated that if all the gill nets used in the Pacific were placed end to end they would reach four times round the equator. When these nets break loose or get lost they continue to fish until they sink to the sea floor. All sorts of creatures can become trapped in them including whales, dolphins and seals. In Alaska nets represent the biggest threat to the Alaskan fur seal. The population has declined by more than half in the last thirty years.

Plastic is virtually indestructible and represents a considerable hazard to marine life. Sea birds have died having got their heads caught in round plastic beer-can holders. Seals have died when they have pushed their snouts into plastic rings and not been able to withdraw them. Turtles have mistaken plastic bags for jelly fish and choked to death. And small plastic pellets are often mistaken for sea creatures and eaten. The stomach of one 11-lb sea turtle was found to contain no less than 2 lb of plastic litter. The plastic makes the creatures feel full and prevents them from eating so they starve to death. Plastics are said to kill up to a million sea birds a year and 100 000 marine mammals.

Plastics cause a problem not just when they are dumped at sea. They have become ubiquitous. In the United States plastics are the fastest growing share of the waste stream – 20 billion lb of plastics are thrown away each year. The average American each year uses 200 lb of plastic, of which 60 lb is for packaging. But now individual States are increasingly insisting that plastics are biodegradable or at least are encouraging recycling of plastics. Some companies, like ICI and Belland in Switzerland, are forecasting large sales of their biodegradable plastics over the next few years. As far as dumping plastics and other waste at sea is concerned new controls on dumping will help reduce the pollution, but as of July 1989 the Annex controlling this sort of dumping had been ratified by only 41 countries whose combined merchant shipping represented only 61 per cent of the world's gross tonnage. And the Annex itself, because it was optional, only came into force in 1989 – 15 years after it had originally been adopted.

CHAPTER TEN

Pesticides and Nitrates

In addition to the accidental contamination of waterways and land with chemicals from waste disposal there have been serious problems in some countries as a result of chemicals which have been purposely used on the land – notably pesticides and fertilisers. Both nitrate-containing fertilisers and chemical pesticides can contaminate water, and pesticides can contaminate food. There has been increasing concern among environmentalists about the effect of nitrates and pesticides on people, and though the concern about nitrates may be overstated, the problem of pesticide use, notably in developing countries, is very worrying.

PESTICIDES

Chemical pesticides have a vital role to play in agriculture. World wide between 20 and 40 per cent of crops are lost each year due to pests like insects, plant diseases and weeds. Even after the crop is harvested another 10–20 per cent can be lost to pests. Chemicals which attack and destroy those pests and control weeds and diseases have provided benefits amounting to three or four times the cost of the pesticides themselves.

In brief there are three types of pesticides: **insecticides** which kill insects, **fungicides** which kill fungi and **herbicides** which kill weeds. Insecticides are divided into four main groups: the organochlorines which include DDT, dieldrin and aldrin; the organophosphorus compounds; the carbamates and the pyrethroids. The fungicides and the herbicides are similarly divided into various chemical groupings.

Until the end of the Second World War farmers relied on such things as crop rotation, the timing of sowing and mechanical weeding to protect crops. There was only a limited use of pesticides. Fungicides were mainly compounds based on sulphur, mercury and copper and herbicides were based on such chemicals as sulphuric acid, copper sulphate and sodium chlorate. It was during and after the Second World War that the first man-made organic insecticides were produced: the organochlorine compounds. In the early fifties products like dieldrin and aldrin were produced, extending control to soil pests. The hormone-type herbicides also appeared at the same time.

Throughout the fifties new types of herbicides, insecticides (including the first contact organophosphorus compounds) and fungicides (including the carbamates) appeared. The use of pesticides became widespread. In the 1960s efforts were made to find more efficient and less toxic pesticides and during this period the number of approved pesticides increased threefold. In the 1970s the rate of introduction of new products slowed, though the use of pesticides on wheat increased dramatically. The use of organochlorine pesticides declined during the early 1970s whereas the use of the organophosphate insecticides increased markedly. Also the use of the newer pyrethroid insecticides became more widespread.

According to the International Alliance for Sustainable Agriculture, in the first five years after the war 20 companies entered the pesticide business and introduced 28 new products. By 1970 the number of companies had more than doubled and the number of new pesticide products had increased to 500. The growth in the market has averaged 12 per cent a year since 1960.

In the United States the use of pesticides in agriculture nearly tripled between 1965 and 1985, in which year farmers applied 390 000 tons of pesticides to the land. Developing countries consume about 20 per cent of world production – and that is increasing. In India in the 1950s farmers used about 2000 tons of pesticides a year. Today they use 80 000 tons a year.

Because of their widespread use traces of pesticides can now be found virtually everywhere, carried to the four corners of the globe through natural air and ocean circulation. Because many of the chemicals are persistent they do not break down chemically in the environment. Of the 70 000 chemicals in common use, only a small percentage are pesticides but they are potentially among the most hazardous because they are deliberately designed to be poisonous to insects and other pests. The fact that they are spread on crops provides them with a direct route into the food we eat. And by contaminating the ground they can be washed into rivers and underground water courses and aquifers where again they have a direct route back to man.

According to 'The Pesticide Poisoning Report' by the International Organization of Consumer Unions, between 400 000 and 2 million people are poisoned each year by pesticides – most in the developing world. Between 10 000 and 40 000 deaths occur each year from pesticide poisoning in farmers and farm workers, many of whom may not have the right equipment for spraying and may not even be able to read the labels to see how the chemicals should be used. In one survey, in 1985, six out of ten farmers in one Brazilian State who had used pesticides had suffered an episode of acute poisoning.

A study by the International Rice Research Institute has found similar problems in farmers in the Philippines who use pesticides. More than half of them had eye and heart abnormalities and 41 per cent had abnormal lungs. The study looked at the comparison between the health of women and men and found that while only 17 per cent of the women had abnormal electrocardiogram readings, the figure rose to 83 per cent for the men. Furthermore, all the men had abnormal chest X-rays, whereas all the women's X-rays were normal.

A report by the Worldwatch Institute published in 1987 which looked at the use of pesticides throughout the world contained some disturbing claims. It highlighted the fact that some pesticides, like DDT, which are banned in the United States and in Europe are still in use in Third World countries. DDT and another pesticide banned in most western countries, BHC (benzene hexachloride), account for three-quarters of the pesticide use in India. When certain foodstuffs were analysed, like cereals, eggs and vegetables, 30 per cent were found to be contaminated with pesticide residues exceeding the tolerance limits set by the World Health Organisation. And samples of breast milk from all 75 women tested were found to contain residues of DDT and BHC. Every day through their mothers' milk, babies were getting on average 21 times the levels considered acceptable. In a similar survey of Nicaraguan women DDT levels in milk were found to be 45 times the World Health Organisation's tolerance limits.

In 1985 because of widespread concern about pesticides the United Nations Food and Agriculture Organisation issued a code of conduct known as the International Code of Conduct on the Distribution and Use of Pesticides. The code is seen as a way to stop irresponsible marketing of pesticides, particularly in Third World countries where controls are less strict than in industrialised countries. The working of this Code of Conduct has been monitored by an organisation called the Pesticide Action Network (PAN) which consists of over 300 non-governmental organisations with an interest in the issue.

PAN's monitoring activity shows mixed results. The code has certainly made countries and companies more aware of the problems. In certain countries tougher controls have been considered and the pesticide industry's international organisation (GIFAP) has agreed to make sure its marketing, labelling and advertising fits in with the code. But its efforts have not satisfied everybody. For example, the Panos Institute – an international policy studies institute promoting greater awareness of sustainable development – has published articles in its magazine which have been critical of the industry. According to one, industry still has to prove itself committed to the spirit of the code by stopping the production of extremely hazardous pesticides and stepping up its research into new safer ones. There are still instances where companies ignore regulations in advertising. Some products are accompanied by attractive advertisements which are misleading and claim products are '100 per cent effective' and 'safe to humans'. It is claimed that some companies sometimes attract customers by giving away free gifts. In other breaches of good practice pesticides may often be put on sale in shops next to foodstuffs, and empty pesticide containers are often used as containers for food or to store grain.

One Kenyan farmer quoted by the Panos Insititute summed up a common problem resulting from this type of indiscriminate use of pesticides: 'One season we opened up in a swampy area without using any chemicals and we obtained 11 250 kilos per hectare. The yellow stem borer was a problem. The damage was not much. Now we are using pesticides. Very costly and we lose our fish. In the beginning it was low input high interest, about 70 per cent . . . Now it is high input low interest – and no fish.' In December 1989 the UN Food and Agriculture Organisation decided on a stricter code of conduct on the use and distribution of pesticides.

Though certain pesticides have been banned in industrialised countries, consumers there are not necessarily completely free from risk because food from Third World countries contaminated with pesticides is still being imported. And, where monitoring exists, it can usually only look at a relatively small percentage of samples. For example, in 1983 the independent Natural Resources Defense Council in the United States monitored Latin American coffee beans sold in New York city markets and found that the four samples tested all contained residues of DDT, BHC and other persistent pesticides.

In 1987 a survey by the US National Research Council suggested that pesticides present in produce grown in the United States could still represent a cancer risk because of contaminants. They estimated that as many as 20 000 cases of cancer a year might be due to contaminated food.

Unfortunately, as the use of pesticides has increased, so has the development of resistant strains of both insect pests and weeds. The result is that in many places the insects and weeds are making a come-back. This tempts the farmer to use heavier doses of chemicals. According to figures quoted by the World Watch Institute, in 1938 scientists knew of just seven insect and mite species which had acquired resistance to pesticides. By 1984 that number had risen to 447 and included most of the world's major pests. Resistance among weeds was almost unknown before 1970. Now at least 48 weed species are resistant to herbicides.

Apart from being a threat to human life, pesticides are also killing animals and birds. In Britain, for example, the toxic effects of organochlorine pesticides (used to protect timbers in houses from wood-boring beetles) have seriously affected the bat population. The population of the greater horseshoe bat has declined by more than 90 per cent in the last century, most of this decline being in the last 30 years. The fungicide pentachlorophenol is also lethal to bats. In the past many species of wildfowl including greylag geese, pink footed geese, brent geese and Bewick's swans have been killed as a result of eating seeds dressed with an insecticide called carbophenothion, which protects the seeds from the wheat bulbfly. Following a number of serious incidents in Scotland between 1971 and 1974, all of which occurred between November and January, the use of this seed dressing for wintering wheat was voluntarily stopped.

But pesticides can affect wild birds in a more indirect way too. The number of wild grey partridges, for example, is now only a third of what it was before the Second World War probably because there are now insufficient insects for the partridge chicks to eat as the weeds which sustained the insects have disappeared.

One technique which has been developed to enable farmers to place less reliance on chemicals is known as Integrated Pest Management or IPM. This involves a combination of making use of natural insect predators, crop rotation and only a minimal use of chemicals to keep pests down. The idea is not necessarily to eradicate pests and weeds, but to reduce them to levels where they do little harm. The experience in those countries which have adopted such techniques shows that they not only work but they also save money.

China is one country which has been spectacularly successful at putting this policy into effect. The Chinese Government have a policy of carefully monitoring and forecasting the spread of pests. Such techniques require a lot of people to make the observations and report back. In many other countries which do not have the 'people power', such methods are not so easy to implement.

The Chinese have also identified natural predators of various insect pests which affect rice, tea, soybeans and other crops. In Guangdong Province, for example, insects which bore into the stems of sugar cane are being kept at bay by the use of a tiny wasp which is a parasite of the eggs of the pest. And at the Dachiao Commune in Jiangsu Province cotton growers plant sorghum between cotton plants to attract the natural enemies of the pests which attack their cotton crops. Together with other techniques, and the minimal use of chemicals, pesticide use has dropped by 90 per cent and costs by 84 per cent. It's a similar story in Brazil and parts of the United States.

In Africa in the seventies and early eighties the mealybug became an increasing threat to a major food crop on which 200 million Africans depend – cassava. The use of pesticides was ruled out because of the difficulties of delivering it to so many farmers in need. Instead a natural predator of the mealybug was found – again a tiny wasp which parasitises the mealybugs' eggs. It has had remarkable effects controlling the mealybug over an area of 65 million hectares in 13 countries.

Non-chemical methods for controlling weeds are also being developed, including the use of fungi, bacteria and other disease-causing agents.

One aspect of policy which has to be addressed if farmers are to be discouraged from continuing to spray pesticides so heavily is subsidies to farmers for pesticides. Some countries have found that by reducing these and phasing them out, money can be freed to pursue Integrated Pollution Management techniques.

It would be wrong to think of pollution by pesticides as only a problem of the developing world. Pesticides contaminate drinking water in many industrialised countries to a greater or lesser extent and in Britain pesticide traces can now be found in drinking water in certain places. In 1988 the Friends of the Earth issued a report which said that drinking water from 298 water sources contained levels of pesticide which exceeded the European Community's Drinking Water Directive. This says that the Maximum Admissible Concentrations of any single pesticide should not exceed one part per billion. In addition Friends of the Earth claimed that the concentration of total pesticides, which should not exceed 5 parts per billion, were exceeded on 76 occasions between July 1985 and June 1987. The problems were confined to the Anglian, North West, Severn-Trent, Thames, Wessex and Yorkshire regions. Friends of the Earth suggest that the absence of reported breaches elsewhere might simply be because breaches had not been looked for there. Among the most commonly found weedkillers were Atrazine and Simazine. It is claimed they are widely used throughout the UK.

The British Agrochemicals Association commented that the EEC's directive takes no account of the wide difference in toxicity which exists between pesticides. The British Government also believes that the directive's Maximum Admissible Concentrations do not always fairly reflect the health impact of individual pesticides on humans – it believes that for some pesticides the MACs are set too low. Nevertheless it does admit that the fact that pesticides are being detected at all in some water sources is a matter for concern, and says that research is being undertaken to assess how and why this is happening. Opponents of the Government claim the Government does not have sufficient information either about the full extent of pesticide contamination in water or the possible health consequences.

Research is now going on to find the best way of getting pesticides out of water. At present the best way is by filtering it through granular activated charcoal, though the charcoal may not remove all pesticides effectively. Some people argue that before you can be sure that all pesticides are removed there has to be a way of knowing exactly what is there in the first place. There are about 300 different pesticides used in Britain today so monitoring water samples is a complicated task.

The identification and proper monitoring of pesticides in water has to be matched by similar monitoring of pesticides in food.* Monitoring in Britain is carried out in three ways: the concentrations of residues in individual groups of foods is monitored – in fruits, vegetables, cereals and so on; surveys are carried out to estimate overall human dietary exposure to pesticides; and surveys are designed to determine the level of pesticides in miscellaneous foodstuffs and in such things as animal feeds and wildlife. Occasionally other studies are organised to investigate specific problems.

In its most recent report, published in 1989, the Working Party on Pesticide Residues outlined its findings of the previous three years. The results make interesting reading and present a fairly reassuring picture. Tests for the environmentally persistent organochlorine pesticides in various food groups showed that on the whole they were found less frequently than in 1981. This is thought to be a reflection of the phasing out of these pesticides in Britain. Similar tests on the levels of organochlorine pesticides in total diet samples show a decline since the 1960s. Tests for the presence of organophosphorus pesticide residues in the total diet showed that the average daily intake for each adult was between 20 and 50 times lower than the recognised acceptable daily intake. It was a similar story for other groups of pesticides.

But while most pesticide concentrations were low there were several occasions where the levels were found to be unexpectedly high and where further efforts at reduction were called for. There were relatively high levels of pesticide residues in certain imported meats from China and high levels of the organochlorine pesticide dieldrin were found in potatoes and eels. Tests on milk and dairy products, though revealing that residues of organochlorine pesticides were in general low, found three samples where the levels did approach maximum limit levels as specified by the European Commission. The Government's Committee on the Toxicity of Chemicals in Food, Consumer Products and the Environment, which considered the working party's report, were concerned about this finding – in particular because milk makes up a large percen-

* The World Health Organisation assumes that people get one per cent of their pesticide intake from drinking water and 99 per cent from food.

tage of the diet of children. They felt that intakes of dieldrin by infants consuming the most contaminated samples of milk could exceed what is considered acceptable and the long-term effects of this are not known. It was suggested in the working party's report that the residues may have been carried over from animal feed.

The Committee also drew attention to a finding by the working party that residues of pentachlorophenol, which is mainly used to treat wood, were found in 32 to 48 per cent of offal, poultry, egg and milk samples analysed in the total diet study. The possible source of this chemical was thought to be wood shavings and bedding used for poultry and its possible use to treat wooden fencing and buildings on farms. The Committee said, 'although we do not consider that the levels detected present any hazard to health, we recommend that all the major sources of pentachlorophenol residues in food are traced and that action is then taken where feasible to remove these sources from the environment of food producing animals.'

Because eels were found to have relatively high levels of organochlorine residues, the 'appropriate bodies' were informed, says the report, and were recommended to take action to reduce environmental contamination by these compounds. The British Government had already announced in 1988 that approvals for the use of aldrin and dieldrin were being withdrawn. Traces of pesticides were also found in wood pigeons and rabbits.

In its conclusions on the working party report, the Committee on Toxicity of Chemicals in Food said that in general it found the report's results reassuring. Where this was not the case they made appropriate recommendations 'not because we consider that there is a hazard to health from the residue levels reported but because we believe it is prudent to keep levels as low as possible'.

Despite the reassuring nature of the report there are still areas of pesticide monitoring and control in Britain which are giving a number of organisations representing both industry and environmentalists cause for concern. They include the delay in reviewing the safety of older pesticides which have been on the market for many years, the delay in the approval of new ones and the shortage of inspectors to ensure enforcement of the regulations.

In 1989 a joint approach to the Ministry of Agriculture was made by a consortium of six organisations, Friends of the Earth, The Pesticides Trust, the National Federation of Women's Institutes, The Green Alliance, the Transport and General Workers' Union and the British Agrochemicals Association. The consortium had called for increased resources so that safety reviews of the older pesticides could be completed by 1992, it wanted more money to reduce the delays on the approval of new pesticides to no more than a year, it suggested that 100 extra inspectors could be recruited and it wanted the frequency of testing for pesticide residues stepped up.

The consortium described the reply they received from the Agriculture Minister, Mr John Gummer, as disappointing. He acknowledged that he was also concerned about the rate of progress on the review of older pesticides but made no firm commitment to increasing resources. And his letter refused to accept that the monitoring of pesticide residues needed a substantial increase in funding. He did, though, acknowledge the need for an independent medically qualified coordinator to be commissioned to collect and evaluate data on pesticide poisoning of human beings. Such a person would report to the Health and Safety Executive.

Though the Minister went some way towards accepting the concerns of the consortium, he fell short of their expectations. But the issue is fluid and public awareness of the concerns is increasing. In the light of that some sort of political initiative may well occur.

Another problem which in some ways is more difficult to control is the interaction or two or more pesticides or herbicides each of which enhances the effect of the other. The phenomenon is known as potentiation. Low doses of two chemicals, each harmless on its own, may result in a lethal mixture if an animal ingests them both.

To cater for those who are suspicious of chemicals in food some farmers have gone into organic farming which means they have to avoid the use of most artificially produced chemicals. Despite its high profile, in Britain only about 1000 out of 110 000 farms are organic. According to the National Farmers Union this is partly because a farmer has to grow food organically for two years before he can call his produce organic, so for this period he faces all the disadvantages of not using chemicals and none of the market advantages of the organic label on his produce. Though organic food is likely to attract an increasing share of the market in the future it is unlikely ever to be more than a minority share. Apart from anything else organic food is more expensive.

NITRATES

As we have seen in the last chapter, nitrates enter the water from agricultural land where they have been used as fertiliser. They are also released from the soil after ploughing. Fears are often expressed about the effect of too much nitrate in the drinking water. Apart from eutrophication there has been much concern that if water rich in nitrates is given to bottle-fed babies under six months old it can lead to what is known as the blue baby syndrome or methaemoglobinaemia. To produce this condition nitrates have to be present in concentrations in excess of 100 milligrammes per litre. This condition is really more of a risk in water which comes from rural wells. In Britain there have only been fourteen cases of the blue baby syndrome in the last 30 years. Unlike the limits on pesticides the limits for the presence of nitrates in drinking water have been set after studies of human epidemiology and as a result the limit is more generous than it would have been if it were based on animal toxicological studies. The present limit is 50 milligrammes per litre and some scientists believe even that is too restrictive.

There have also been suggestions in the past that high concentrations of nitrates might lead to cancer as a result of the formation of nitrosamines in the gut. Nitrosamines are known to be carcinogenic. But there is no evidence linking nitrate levels with cancer in Britain. In fact studies have shown that in areas where nitrate levels are relatively high the incidence of stomach cancer is actually lower than expected.

Even if water is high in nitrates it is possible to blend it with water low in nitrates from a different source so that the result at the tap is water within the limits, though this can be an expensive business. Nitrates can also be taken out of water by the use of ion exchange resins or by biological denitrification – though these methods are very expensive.

In Britain the worst affected areas for nitrate in water are the Anglian and Severn-Trent areas but there are also problems in some parts of the Southern, Thames and Yorkshire areas. The most severely contaminated water is in the Jurassic limestone of Central Lincolnshire where some water companies have had to take some boreholes out of supply or use them intermittently. Sometimes the water from these boreholes is blended with low nitrate water. Recently there has been a decline in nitrate concentrations in some of these boreholes, possibly due to a change in agriculture from spring to winter cereals.

The July 1988 report of the House of Commons Select Committee on the Environment stressed that the current limit of 50 milligrammes of nitrate per litre of water was based on the science of a decade ago and it urged the Government to ask the European Commission to revise its standards. It doubted that any code of good agricultural practice would achieve the necessary reductions in the application of nitrogen fertiliser and suggested that trial protection zones should be set up in selected catchment areas in which the use of nitrogen fertilsers could be regulated or prohibited.

CHAPTER ELEVEN

Water, Water Everywhere . . . ?

So far we have looked at the problems of pesticides and nitrate contamination of water. But many other sorts of wastes are being discharged into rivers and seeping into water courses, and chemical contamination of drinking water is a broader issue. It is true though that on a global scale chemical contamination is nothing like as important as contamination by water-borne organisms. In the Third World, each day, 25 000 people die from water-borne diseases.

Drinking water in European Community Countries is governed by a European Commission Directive on water signed in 1980. It contains a whole list of 66 standards covering everything from toxic chemicals to the appearance of the water. Some scientists and water experts now say that before it was signed the full implications were not taken into account. As we have seen, the result has been that several European countries, Britain included, have had trouble keeping strictly to all the standards laid down. The British Government believes that in some ways the UK regulations made under the 1989 Water Act actually go beyond the European Directive – the standards for lead contamination are tighter in Britain, for example. But there has been considerable controversy over the Government's decision to allow some water companies dispensation not to meet certain strict standards. The Government says dispensation has been allowed where there is no danger to health and where the failure to meet the standards results from natural causes. But according to environmentalists some of these dispensations granted on the grounds that the contamination is present as a result of nature are not justified. Friends of the Earth make the point that the majority of lowland reservoirs are rich in nutrients from sewage and the run-off from highly fertilised arable land and that is not due to nature. Groundwater may also be contaminated as a result of oil spillages – again a man-made problem and not something which should exempt the water authority from having to conform to legal limits.

Chemicals in water can affect its taste, smell and appearance. All water contains chemicals in small amounts. Indeed it is the chemicals which give it its 'natural' taste. Pure H_2O with all the chemicals removed tastes very odd indeed. But the presence of certain organic chemicals, like the pesticides considered in the last chapter, in large enough concentrations to present a health threat is quite a different matter.

In the 1960s a detection technique called gas chromatography was developed which was able to reveal that water did in fact contain minute quantities of organic compounds whose presence had not been suspected. Amongst these were chemicals which, in higher concentrations, would undoubtedly be harmful. The problem is, there is uncertainty about the long-term effects of very low concentrations – several parts per thousand million. If low concentrations like this are affecting health any effect would take a very long time to develop and therefore it would not be easy to detect the effects by a study of the patterns of disease: the technique known as epidemiology. Instead it is possible to estimate the relative hazard of various chemicals by looking at their effects on animals at higher concentrations. Toxicological studies, as they are called, seek to estimate how poisonous a chemical is by giving it to certain species of animals at various concentrations and seeing what happens. The World Health Organisation uses animal studies like this to set a level of any particular chemical which is judged to be 'acceptable' for humans. It is known as the Acceptable Daily Intake or ADI. The ADI is set by finding the level of a chemical which has no adverse health effects on two different species of animal and then reducing the level in factors of ten. That means that for some chemicals the acceptable daily intake for man may be a thousand times less than the amount needed to cause any harm to animals.

LEAD

Lead may be present in water as a result of contamination from the lead pipes through which the water runs, particularly in older houses and particularly in soft water areas. Soft water is more likely to dissolve lead from pipes. The current European Commission standard for lead is 50 millionths of a gramme per litre. Some scientists think it should be lower. It is now generally accepted that low levels of lead have a neurotoxic effect and there is evidence it can adversely affect children's intelligence. Exposure to lead from water can be reduced by treating the water so it is not so corrosive to the lead pipes. But the best way of getting rid of lead in water is to replace lead pipes with copper or plastic pipes. In Britain about 45 per cent of houses still have lead pipes and replacing them would be very expensive.

ALUMINIUM, IRON, ZINC

The levels of aluminium, iron and zinc are controlled in water supplies not so much on health grounds but because they affect the taste, smell or appearance.

MICROBIOLOGICAL STANDARDS

Whatever risk chemical contamination of water might pose to human health it is relatively less significant than the risk of micro-organisms in water. The first demonstration that water contaminated with bacteria could cause disease came in 1854 when an outbreak of cholera in London was stopped when John Snow removed the handle of the water pump in Broad Street. The key to healthy water is keeping human waste out of the water supply. Faecal contaminants can in theory find their way into water supplies where, for example, inlets to water supplies are downstream of sewage outlets. There is also a risk of contamination of reservoirs from cattle and birds.

Water is treated with chlorine to kill the germs it might contain. In Britain in the past 50 years there have been few outbreaks of typhoid or paratyphoid and the incidence of dysentery has fallen. On the other hand some organisms are more resistant to chlorine. Viral gastroenteritis has become more common and cryptosporidium is sufficiently resistant now for the security of the water supply to depend more heavily on filtration. In 1989 there were outbreaks of disease caused by cryptosporidium in Oxford and Swindon.

Germs from the gut which can cause disease if present in water are sometimes difficult to detect and so scientists use 'marker' organisms to test the water. They look for harmless bacteria called coliforms which are present in large numbers in the human gut. They decay at about the same rate as the disease-causing germs and are easier to detect. (Some coliforms called thermocoliforms can survive at temperatures higher than 37 degrees.) If scientists discover these are present then they can assume that the water has been contaminated with human or animal excreta and there is a possibility that disease-causing germs may also be present.

The European Community lays down limits for the number of coliforms and thermocoliforms which may be present. In a 100 ml sample of tap water there should be none.

Another problem which became apparent in Britain in the hot summer of 1989 was the growth of potentially highly toxic blue-green algae in reservoirs. The contamination came to light publicly after sheep and dogs had died after drinking water from the Rutland Water reservoir owned by Anglian Water. The National Rivers Authority warned the owners of certain reservoirs in Buckinghamshire, Cambridgeshire, Essex, Leicestershire, Lincolnshire, Suffolk and Norfolk to consider banning recreational activities such as wind surfing, fishing and sailing, to advise the public not to swim, paddle or handle the water, to keep livestock away from it and not to let their dogs swim in it. Environmental organisations blamed the increase in nutrients in the water from agriculture for the blooms of poisonous algae.

* * *

WHERE DOES WATER COME FROM?

In Britain on average 70 per cent of tap water comes from rivers, lakes, streams and reservoirs. In its natural state river water should be drinkable with only the minimum treatment. But as we have seen, rivers are used as a dumping ground for all sorts of wastes from homes and industry and are also contaminated by agriculture. Despite this, 90 per cent of Britain's rivers are said to be safe for use as drinking water.

But that leaves 10 per cent of rivers which are not safe and environmental organisations have been tenacious in pursuing industry, water companies and Government in order to get improvements in river water quality. The Government's own digest of Environmental Protection and Water Statistics published in 1989 shows that between 1985 and 1987 there was a decline in the length of good quality rivers (by 172 km). And the number of water pollution incidents rose by 9 per cent in 1987 to 23 253. That compares with a level of only 12 500 in 1980. Industry was the main cause of the incidents – it was responsible for 37 per cent. Nineteen per cent of the incidents were caused by farm pollution and 20 per cent were due to pollution by sewage. Only 1.5 per cent of the incidents resulted in prosecution and even that low figure was more than the previous year. Sixty per cent of those prosecutions followed incidents of pollution by agriculture and the Government report makes it clear that agricultural contamination can be particularly serious because it may involve discharges of large amounts of organic matter which can cause great damage. Silage effluent is up to 200 times more polluting than untreated sewage whilst cattle slurry is 100 times as strong.

The official figures do not only relay the bad news. They also show, for example, that between 1985 and 1987 the total length of bad quality rivers has declined – by 206 km. In 1989 in a report called Poison in the System, Greenpeace reproduced some of the Government figures and drew other less flattering comparisons. Their report claimed that in 1986–7, 4345 km of rivers were effectively biologically dead – 10 per cent of all rivers and 3 per cent more than in 1980. Rivers classified as good quality rivers declined from 69 per cent in 1980 to 67 per cent in 1986–7.

The Greenpeace study also revealed how difficult it is for the public to find out exactly what is being discharged because industry not only discharges directly into rivers, it can also discharge waste into the sewage system. The problem is that sewage works are not designed to deal with some of the wastes they are getting from industry so the waste contaminates sewage sludge and can also be discharged to rivers.

About a third of Britain's water comes from under the ground, eighty per cent of it from two main aquifers. Where aquifers are exposed they may be vulnerable to pollution, but they are less vulnerable when they are sealed above and below by impermeable strata such as clay. The two things which pose the greatest threat to aquifers are pollution by agriculture and waste disposal in landfill sites.

The contamination by pesticides and by solvents from industry is potentially a serious problem because the permitted level of these substances are so low that a small amount of contaminant can pollute a large amount of water. And what makes it more difficult to control is that nobody fully understands how organic pollutants migrate in underground aquifers.

WATER SHORTAGES

Apart from the contamination of water, there is another aspect of water supply which is just as important. We have not only to worry, on a global scale, about whether the water itself is healthy but whether it is there at all.

For decades all over the world water has been extracted with little thought given to whether it would ever run out. Yet recently, for the first time, we are having to face the consequences. Fresh water is a scarce resource and not only in the obviously dry countries. It is estimated that within 15 years in many areas of the world demand for water will exceed supply. In the United States water tables are falling beneath a quarter of the areas where crops are irrigated. Of particular importance is what is known as the Ogallala Aquifer which stretches from

EXAMPLES OF WATER SHORTAGES IN CERTAIN COUNTRIES AND REGIONS

North and East Africa:	Ten countries likely to experience severe water stress by the year 2000. Egypt is already near its limits and could lose vital supplies from the Nile as upper-basin countries develop the river's headwaters.
China:	Fifty cities facing acute shortages. Water tables below Beijing are dropping 1–2 metres per year. Farmers in Beijing region could lose 30–40 per cent of their supplies to domestic and industrial uses.
India:	Tens of thousands of villages now face shortages. Plans to divert water from Bramaputra River have heightened Bangladesh's fear of shortages; large portions of New Delhi have water only a few hours a day.
Mexico:	40 per cent more water is being pumped from the valley where Mexico is sited than is being recharged, causing land to subside. There are few options to import more water.
Middle East:	Israel, Jordan and the West Bank are expected to be using all renewable sources by 1995 so shortages are imminent. Syria could lose vital supplies when Turkey's huge Ataturk Dam comes on-line in 1992.
Soviet Union:	Depletion of water has caused the Aral sea to drop by two-thirds since 1960. Tens of thousands have left the area.
United States:	One fifth of total irrigated area is watered by excessive pumping of groundwater. About half of the western rivers are over-appropriated. To augment supplies cities are buying farmers water rights.

Source: *State of the World 1990, A Worldwatch Institute Report on Progress Toward a Sustainable Society.* © Worldwatch Inst.

southern Dakota to northwestern Texas. This consists of virtually non-renewable ground water and it is diminishing fast. Water is being pumped from the aquifer to irrigate the Texas plains where 40 per cent of the country's grain-fed beef are raised. The pumping is responsible for 70 per cent of the depletion of the aquifer.

In the Soviet Union expanding crop production in the central Asian republics and Southern Kazakhstan is threatened by a water shortage. In the years when the weather is very dry virtually all the flow of the two main rivers in the area is already used. In the North Plain of China water tables are dropping at the rate of 2 metres a year. The Washington-based Worldwatch Institute forecasts that this part of China as well as nearly all North Africa, certain places in India, much of the Middle East and parts of the western United States could be entering a period of chronic water shorage in the 1990s. A Worldwatch report on water published in December 1989 summed it up: 'With the world projected to add some 96 million people per year during the nineties, efforts to improve water productivity can only be stopgaps.

'The imbalance between human demands and local water supplies will worsen without major efforts to slow population growth and eliminate wasteful practices.

'In a country such as Egypt, where the population leaps by one million every eight months, modernizing irrigation systems is simply not enough. And in the western United States, using scarce water to irrigate hay for cattle may not be possible for much longer.' The report ends, 'Over time, spreading water scarcity will lead to food shortages and rising food prices. These and other consequences of overstepping water's limits are inescapable; for much of the world, time to avert them is rapidly running out.'

The fall in water tables not only threatens agriculture. It is also a threat to the stability of the land itself. Overpumping has led to land subsidence due to compaction in Mexico City, Beijing, California's Central Valley and the Texas Coast.

Bad management of water supplies can lead to far more water than is needed being used in fields. Even fresh water contains some salt – typically some 200 to 500 parts per million. When the water evaporates it leaves behind these salts causing salinisation. Unless this salt is flushed away it builds up eventually making the land unusable. If about 10 000 cubic metres of water a year are added to each hectare of irrigated land then between 2 and 5 tons of salt is added to the soil each year.

Perhaps one of the most dramatic demonstrations of the shortsightedness of using water unwisely happened at the Aral Sea in Soviet Central Asia. So much water for irrigation has been taken out of rivers supplying the Aral Sea that the area of land covered by the sea has shrunk by more than two-thirds since 1960 and salinity has tripled. Around the perimeter of what was once sea, stand fishing villages now robbed of their livelihoods, their boats stranded miles from the water, the bed of the sea now cracked hardened mud. All native fish species have disappeared.

As Third World countries increase their demands for water to supply expanding populations it is essential that water is husbanded more efficiently than it has been up till now. This is already happening in many areas, with 'drip' irrigation being used – a technique in which water is trickled or dripped at or beneath the surface direct to the roots of the plants. This avoids flooding and waste of water through evaporation.

It is not only agricultural use which is draining water from lakes and rivers. In the city of Los Angeles, environmentalists scored a major victory in late 1989 when they legally prevented the city's water department from diverting any more water from Lake Mono – an alkaline lake high in the Sierra Nevada mountains. Los Angeles has been diverting water from the lake since 1941, at which time the lake was 41 feet higher than it is at present. There had been fears that at the existing rates of consumption, by early in the next century the lake's ecosystem could collapse. The lake is an important breeding ground for thousands of rare and threatened birds and the environmental movement had been fighting for some time to protect it. The judgement also insisted that the city returned some of the water it had already drained into reservoirs to raise the water level of the lake by a foot and a half.

Water is also needed for industry, and increasingly industry is learning to re-use and recycle water. In the United States, for example, in 1978 manufacturing industries used 49 billion cubic metres of water. Each cubic metre was re-used on average 3.42 times before being discharged, saving the need to withdraw some 120 billion cubic metres from the nation's water supplies.

More than 80 per cent of water used in manufacturing is used in just four industries: paper, chemicals, petroleum and primary metals. Some industries have made their use of water much more efficient – paper mills for example now re-use water more than seven times before discharging it. As techniques improve recycling rates are going up. By the year 2000 it is estimated the recycling rates in paper making and in the manufacture of primary metals is expected to rise to 12, in chemicals to 28, and in the petroleum industry to more than 30.

CHAPTER TWELVE

Saving the Plants and Animals

It is perhaps still not too late to reverse much of the impact mankind is having on the environment even if, as with the ozone hole and the greenhouse effect, it may take many decades, even centuries. But there is one aspect of environmental degradation which it is already impossible to reverse: many unique species of plants and animals are now extinct. Once a plant or animal species becomes extinct there is absolutely nothing that can be done to bring it back. It is gone forever, and with it part of the diversity of creation which makes the natural world such a beautiful place. Gone too are any possible benefits which might have been available to humankind had scientists been able to investigate the properties of some of the species now lost. What causes the extinctions is primarily the destruction of habitats but also the exploitation of animals often illegally and often for frivolous reasons.

One of the prime concerns is the destruction of tropical forests. In 1950 tropical forests covered nearly a quarter of the world's landmass. Today they cover less than 7 per cent. During 1989 alone an estimated 142 200 square kilometres of tropical forests were lost – nearly 2 per cent of what was left. It is an area equal to 60 per cent of the size of Great Britain or roughly the same size as Florida. Put another way it means that an area the size of six football pitches is being destroyed every minute. Some believe that even these figures are an underestimate. Satellite photographs suggest that forests covering up to 20 million hectares a year are being destroyed – an area seven times the size of Belgium.

The ten countries worst affected are losing forests at the rate of 4000 square kilometres or more a year. They include Burma, Colombia, Mexico and Thailand. But it is particularly in Brazil, Indonesia and Zaire, which contain half the world's forests, that destruction is greatest, amounting to 66 000 square kilometres a year. In Brazil alone destruction amounts to 50 000 square kilometres a year.

If the current rate of destruction were to continue, in 38 years' time there would be no forests left at all outside protected areas. That is unlikely to happen though some countries are certain to lose all their forests.

One consequence of the destructions of forests is the release of carbon dioxide gas – the main cause of the greenhouse effect. Because the trees are no longer there they cannot act as a store for carbon dioxide, removing it from the atmosphere through photosynthesis. The burning of the dense, so-called 'closed' forests in 1989 released 1.4 billion tonnes of carbon as carbon dioxide into the atmosphere – 75 per cent more than in 1979 and 30 per cent of the amount going into the atmosphere from the burning of fossil fuels. The relative contribution of tropical forest burning to the greenhouse effect is increasing all the time and some estimates suggest that by early next century the burning could be releasing up to 5 billion tonnes of carbon into the atmosphere. After that there would be a decline simply for the reason that there would be very little tropical forest left to burn.

Tropical forests contain over half the world's plant and animal species. And with the destruction that has taken place in recent decades it is a matter of conjecture how many irreplaceable plants and animals have been lost forever. The fate of the Amazon rain forest will serve to illustrate the problem facing all forests.

THE AMAZON RAIN FOREST

The Amazon rain forest covers seven countries and is the largest and most complex ecosystem of all. It is more vital to the health of the planet than any other. It is an 'endless' sea of vegetation, through the centre of which runs the world's largest river – the Amazon.

A recent book on Amazonia* gives a fascinating insight into the importance of the Amazon and the impact mankind is making on the area. The river drains an area of more than 6 million square kilometres. The river itself has no fewer than 10 000 tributaries and flows for 6762 kilometres. It contains two-thirds of the world's 'free' fresh water and delivers to the sea as much water in one day as the Thames delivers in a year. For some way upstream it is wider than the English Channel – so wide in fact that a boat in the middle can see neither shore. The mouth of the river is 240 kilometres across. The source of the river is the Peruvian Andes at which point the river is only about 90 kilometres from the Pacific coast. But it flows east dropping steeply for the first 1000 kilometres. The river then becomes very flat and for its remaining journey of some 5800 kilometres the total drop in level is a mere 300 metres. This means among other things that it is tidal up to 600 kilometres inland.

Apart from being tidal, the river is prone to flooding when the snows in the Andes melt and the extensive flood plains extend up to 100 kilometres from the river. The result is a huge cauldron of biological diversity so rich in species of both plants and animals it almost defies imagination. Many of the trees and fish live in a symbiotic relationship – each helping the other – the fish need the seeds from the trees for food, the trees need the fish to distribute and prepare the seeds for germination. Fish have adapted well to their habitat. Some of the fish have jaws and teeth which have evolved to allow them to crack even tough brazil nuts.

The sheer diversity of species is astonishing. There are 2000 known species of fish in the Amazon – and possibly another 1000 which have yet to be discovered. One-fifth of the world's bird species live in the forest and there are possibly several million animal species, of which many are insects. It is thought that there could be up to 10 000 new plant species waiting to be discovered.

The tragedy is that many of the plants and animals are

* *Rainforests, Land Use Options for Amazonia.*

extremely rare, the particular species in one hectare being different from those in a neighbouring hectare. That means that every time a hectare of forest disappears to make way for a cattle ranch or some other development, priceless genetic material disappears for ever. It is estimated that in the Amazon one species becomes extinct every day. (World wide as many as 20 to 50 species are being lost each day.) We shall never know how useful they might have been to man had they been investigated.

Much of the land is being cleared to produce food, but the irony is that the Amazon river itself contains a huge amount of food which man could exploit without having to destroy the forest. It could produce a quarter of a million tonnes of fish a year, equivalent to the beef protein which might be reared on 25 000 square kilometres of cattle ranch.

It is not surprising that an ecological system this large has a profound effect on the climate, not just locally, but throughout the whole region. The water is constantly circulated. It is transpired by plants and evaporates, and then falls as rain which percolates through the vegetation to begin the cycle over again. As the forest is reduced so the climate will inevitably change, becoming drier. This will probably affect vast areas of land to the south where agriculture is such a vital part of the economy.

But possibly more important than all this is the fact that the destruction of the Amazon forest could on its own have a devastating effect on the climate of the entire world through the greenhouse effect. It is this possibility which has concentrated the minds of so many of the world's governments. And which, at last, seems to be encouraging them to bring pressure to bear on the Brazilian Government to slow down the forest's destruction.

What has stimulated deforestation in the Amazon is the construction of new roads. The National Institute for Research in the Amazon has said that the deforestation which has occurred in Rondonia, Mato Grosso and southern Para is likely to spread to other areas as access improves. Roraima is now being extensively deforested as a result of migration from Rondonia and is the most rapidly growing frontier in Amazonia.

An ambitious road building programme in Amazonia led to a doubling of its population between 1960 and 1980. The road building and government subsidies helped to open up the interior to cattle ranchers. But cattle ranching is inefficient because the soil is of poor quality and notably lacking in phosphorus. To enable pasture to grow properly the soil should contain at least 100 parts per million of phosphorus. When the forest is newly burnt, the ash from the vegetation provides plenty of phosphorus, but this is gradually depleted so that five years after the burning of the trees the level of phosphorus is equal to only about two parts per million. Certain weeds are well adapted to low phosphorus levels and they invade pastures making them virtually unusable. Apart from the lack of nutrients, the soil becomes compacted and eroded. The result of this is that it becomes impossible to sustain a ranch. For many of those who develop the forest that does not matter too much because by establishing a ranch it allows them to claim ownership of the land. They can then sell it when the value of the land rises. This happens as the area becomes more accessible or when others want to move in to exploit minerals or other resources. Dam building projects have also resulted in the destruction of areas of forests.

In 1981 the World Bank lent Brazil money to develop the northwest of the country in what is called the Polonoroeste Programme. The idea was to steer migration away from the fragile areas including those areas where the Amerindians live and encourage sustainable agriculture elsewhere. But since the programme began many environmental organisations have criticised the scheme and the World Bank saying the programme has caused the highest rate of deforestation in the Amazon. They fear it is likely to lead to an area the size of Britain being deforested within a decade. Nearly half the aid has been spent on a main road which has simply encouraged higher rates of migration into the area. They claim the Bank has lost control over the project. One prominent critic has suggested that instead of encouraging workers from the south to come to an area where soil is poor the Bank should be encouraging them to stay in the south and northeast.

In 1985 the World Bank, in collaboration with Brazil's Government, suspended funds for the Polonoroeste Programme admitting that there had been some problems. A remedial action plan was discussed and implemented beginning in 1986. In its defence the Bank said that the area was already suffering from environmental problems even before the Bank became involved.

The impression is sometimes held that it is the rest of the world which is primarily concerned with conserving the Amazon forests while Brazil's Government and those of surrounding countries are intent solely on exploitation and destruction. But that is not necessarily true. There have been three key policy initiatives which have highlighted the Government's apparent enthusiasm for the proper husbandry of resources:

1 **The Amazon Pact** was signed in 1978 by Brazil, Bolivia, Colombia, Ecuador, Guyana, Peru, Surinam and Venezuela. Though this was primarily a treaty for cooperation in development, Article VII refers to the need 'for the exploitation of the flora and fauna of the Amazon to be rationally planned so as to maintain the ecological balance in the region and preserve species'.

2 **The Forest Policy Report** was drawn up by an official high level interministerial commission of the Brazilian Government. It followed internal concern when, in 1978, the Government proposed to grant unrestricted timber rights to multinational logging companies in huge tracts of the rain forest. The report was finished in 1982 but has remained unpublished. It is thought to recommend the preservation of much of the Amazonian rain forest but it was opposed by some in the government and by business interests who believe it was the work of radical environmentalists. One optimistic sign which has encouraged many environmentalists is the appointment of Jose Lutzemburger as Minister for the Environment in Brazil's new Collor Government. He is the most famous forest ecologist and activist in Brazil.

3 There has been much grass roots **opposition from the rubber tappers** who have been living in the forests for generations.

They have flexed their muscles, claiming that they are the one group to have been excluded from consideration when schemes to develop the forests are discussed. They have been campaigning for measures which would preserve their communities – and their campaigns have been partially successful. By October 1988, 12 reserves covering 2 million hectares had been established in five Amazonian states for this form of activity but there is no guarantee the areas will remain safe.

The rubber tappers' fight has not been easy. In December 1988 the leader of the National Rubber Tappers Association, Francisco Mendes, was shot and killed, it is believed, at the behest of those with powerful vested interests. He was the founder of the Union of Forest Peoples – an alliance of Indians and rubber tappers and had made himself unpopular with some sections of the community by denouncing the involvement of large American, Dutch and Japanese companies in forest destruction. He was instrumental in persuading the World Bank and the Inter-American Development Bank not to finance a big road scheme until environmental protection measures had been taken.

Despite the good intentions, the burning and destruction goes on, with 'guidelines' and 'safeguards' being constantly flouted. In 1989 the amount of clearance by burning was in fact less than the previous two years but this was mainly because the rainy season came early. Between 10 and 12 per cent of the forests have already been destroyed.

One important initiative to help control the destruction of the world's forests is the Tropical Forest Action Plan which was launched in 1985 by the World Bank, the UN Food and Agriculture Organisation, the UN Development Programme, and the World Resources Institute. The Plan paved the way for a country by country assessment of forests to identify what technical assistance is needed and what should be the investment priorities in forestry. The assessments include studies on forestry protection and plantation, logging, the use of wood for fuels, conservation of ecosystems and forestry research.

The idea is that once such a review is completed, round table discussions should take place to establish a plan of action. The discussion is intended to include representatives of the affected communities and environment organisations. But many of the non-government organisations complain that local organisations and environmental groups have had little involvement and that all draft plans are confidential and inaccessible to them. The organisations also complain that all Tropical Forest Action Plans have been too inclined to blame small peasant farmers for the problems, have been too orientated towards logging and commercial plantations and have not concentrated enough on such things as conservation, respect for tribal rights and community-based forestry. Friends of the Earth claims the Action Plan is encouraging the very processes which destroy forests.

It is true that in terms of acreage destroyed it is the small-scale farmer who is responsible for more destruction than the commercial loggers and cattle ranchers put together. But it is also the small-scale farmer who is least able to stop clearing the forest because he is the poorest of the three. The individual Action Plans often advocate the setting up of commercial plantations of fast-growing species which according to environmentalists may well be damaging to the environment. By November 1989 the Tropical Forest Action Plan had resulted in forestry reviews in 32 countries and about six countries had had round table discussions. By February 1990, 12 complete plans had been produced.

There has also been progress on another front. Recently the Colombian Government has granted rights to its indigenous forest people over some 18 million hectares – two-thirds of the Colombian Amazonia and equivalent to the size of Britain. The message is clearly getting through.

So far we have concentrated on the problems of deforestation in the Amazon and the dire consequences for the preservation of genetic diversity and rare plants and animals. But many other areas of the world face similar problems. The forests of the **Ivory Coast** and of **Nigeria**, for example, are now virtually gone, after forty years of destruction.

In **Thailand** satellite images show that forests are disappearing at the rate of more than 8 per cent a year – only a quarter of the land remains forested. The destruction of the forests has led to landslides, one of which, in November 1989, left 400 dead.

In **Indonesia** a plan to clear an area of forest half the size of Wales so that fast-growing eucalyptus trees could be planted caused a major outcry. The US giant Scott Paper had hoped, in partnership with another company, to spend $654 million clearing the tropical forest in Irian Jaya and using the trees to make pulp for use in a wide range of paper products. The company in the end did not go ahead with the plan because it said it had discovered it could meet its anticipated needs for pulp from other sources. Friends of the Earth believes that its threat to organise a boycott of Scott's products was instrumental in the company's decision to pull out of the Indonesian forest.

In some of the smaller countries, where destruction has been continuing, there is an increasing recognition that the destruction has to stop. In the **Philippines** forests constitute less than a quarter of a per cent of the world's forests and less than half a per cent of the total amount of tropical forests but President Corazon Aquino has said that in her term of office she is determined to achieve transition from a country where forests are being depleted to a country where the forests increase: 'What we do with our forests may not significantly worsen the global picture but we do feel the obligation that everyone should contribute their share.'

The movements to help preserve rain forests have taken many forms. In **Australia** the Building Workers' Industrial Union has banned the use of new rain forest timbers, much of which is used to build moulds for pouring concrete. They have been supported by waterfront unions who have organised rolling strikes aimed at ships carrying tropical timbers. There have been international boycotts of Japanese traders who import tropical timbers – 92 per cent of them from Sarawak.

There have been suggestions that in order to discourage the cutting down of tropical timbers, wood products should be clearly labelled so that consumers can tell whether the wood has come from a properly managed sustainable source. At present less than one-eighth of a per cent of timber comes from such a source.

The International Tropical Timber Organisation has come under considerable pressure to take greater care of the resource on which it depends, notably from the World Wide Fund for Nature. In a policy document in October 1989 it said that by 1995 the trade should have based itself on sustainable management techniques. It also urged the timber industry to recognise the land rights of the Kayapo Indians of Brazil, to establish a network of model forest management projects and to recognise the importance of conservation.

In some countries effective pressure to plant trees has come from a few dedicated individuals. In **Africa**, for example, a scheme devised by Wangari Maathai, a former professor of anatomy at Nairobi University, to plant trees was laughed out of court fourteen years ago when she tried to promote it. Since then she has persuaded thousands of communities throughout Kenya to plant 10 million trees – more even than Kenya's forestry department. She argued successfully that trees met all manner of basic needs: firewood, fodder, fruits, honey, timber, medicines, building and fencing materials as well as protecting the soil from erosion and maintaining the ecology.

Planting trees can also help slow down the greenhouse effect as we have discussed. One million square kilometres of trees absorb one billion tonnes of carbon during their growing period. Assuming that the net build-up of carbon in the atmosphere amounts to some 4 billion tonnes a year that means that to 'neutralise' this build-up, 4 million square kilometres of forests would need to be planted. That compares with current tree planting in the humid tropics of 10 000 square kilometres a year. The cost would be huge – a ten-year programme of tree planting would cost ten times the current annual expenditure of the Tropical Forest Action Plan. That may be too ambitious, but even a small amount would help.

One important stimulus which is likely to lead to more action will come from the Intergovernmental Panel on Climate Change which will highlight the role of the forests in maintaining the existing climate. A recent report by Friends of the Earth recommended that there should be better assessment of deforestation, particularly by means of remote sensing satellites, so that fresh information can be ready for the Second World Climate Conference in Geneva in December 1990. One important study is being undertaken by the Food and Agriculture Organisation in conjunction with the United Nations Environment Programme. Their report on deforestation is due to be ready by 1992.

As we said at the beginning of this chapter, tropical forests contain a rich diversity of living species, both animals and plants, and as the forests are destroyed so too are the species. Because some of the species are so rare no one will ever know exactly how many species as yet undiscovered are being eliminated. Apart from the fact that the plants and animals are unique and therefore beautiful and awe-inspiring in themselves, on a more pragmatic level they may also be useful to mankind.

More than 120 medicines are extracted from plants. About three-quarters of them were discovered by chemists who were looking for the active ingredient in plants already used for medicinal purposes by indigenous people. The 120 drugs come from just 90 species of plants. In view of the fact that there are probably some 250 000 higher plants in the world, scientists are confident that there are many more useful drugs yet to be discovered.

The World Wide Fund for Nature has listed more than 1000 plants used by the Indians of the South American rain forest for various purposes including food and medicines. For 3000 years, for example, the powdered snakeroot plant had been used in India to treat 'mental agitation'. That is the origin of the tranquilliser, reserpine, which was isolated from the plant in the 1950s. Aspirin originally was based on extracts of the willow bark.

Perhaps one of the most famous examples of plants yielding drugs is the Madagascan periwinkle plant, from which two major anti-cancer drugs are derived. They have generated sales worth $170 million a year.

And discoveries are still being made. A recent pamphlet from the World Wide Fund for Nature describes how villagers in Ethiopia living downstream from a communal washing site were found to be free of bilharzia. It was because the clothes were being washed with dried wild soapberries which produced a substance which was killing the snails carrying bilharzia.

Plants have also provided a useful source of pesticides – their natural way of resisting disease having been tapped and exploited by man. The most famous example of this is probably the wild African chrysanthemum which produces pyrethrins which are valuable insecticides.

Genetic variety is essential if useful crop plants are to be protected against new pests and diseases and maintain their vigour. Useful genes can be bred into new crop plants from existing wild strains. This 'topping up' with fresh genetic characteristics is said to have improved steadily the productivity of main crops in the United States to the tune of $1 billion.

The World Wide Fund for Nature in its efforts to stimulate interest in saving plants gives a number of examples where wild strains have proved invaluable in safeguarding crops. A wild species of rice found in Central India is the only known source of resistance to what is known as the grassy stunt virus, and has been used to breed a strain of rice known as 'IR 36' which is now the world's most widely grown rice variety. There are a number of wheat species which have had bred into them resistance to a range of diseases as well as characteristics which enable them to tolerate drought and both hot and cold climates.

Between 1870 and 1900 virtually all Europe's grape vines were destroyed by an insect pest. The grapes which now grow there are derived from rootstocks from a combination of three North American wild grape varieties.

The production of sugar from sugar cane has doubled since characteristics from wild species were bred into cultivated varieties. The breeding of more productive sunflowers using wild species has led to a 20 per cent increase in yield. And there are more promising developments to come. Maize has to be planted each year, but a species of maize which is perennial has been discovered in western Mexico. Although it has not yet been exploited, its discovery has been described as the botanical find of the century. Apart from being perennial it may well be resistant to many diseases which can strike the crop variety. It is also resistant to cold and damp and may therefore enable maize to be grown in areas which up till now have been unsuitable. Then there is the buffalo gourd which can live as long as forty years and which can supply a variety of foods like starch and oil. It can tolerate extreme drought and is therefore an ideal plant for arid lands. When it has been planted experimentally in deserts in America it has produced a rich harvest.

In the 1960s, wheat crops in the United States were saved from an epidemic of stripe rust by genes derived from a wild species of wheat from Turkey. The United States Department of Agriculture estimate that this one change was worth $50 million a year in increased yields in America.

There are hopes that new strains of plants can be developed which can grow on soils which have a lot of salt in them. As a considerable proportion of the earth's surface is unsuitable for agriculture because of salinity, the development of salt-tolerant plants could revolutionise agriculture in many areas.

The World Wide Fund for Nature draws attention to a potentially serious problem, however, in exploiting genetic resources. Most of the world's supply of 'wild' species from which useful characteristics may be gleaned grow in Third World countries in undeveloped forest areas. It may be thought unjust that once these characteristics are exploited by the developed world and incorporated into new strains of useful crop plants the Third World has to buy them back. They get no income from the genetic resources which were theirs in the first place. Some countries have already taken action to prevent others from cashing in on the natural resources of the plant kingdom. Ethiopia, for example, prevents certain plants leaving the country and Mexico has banned the collection of most of its wild plants.

It is not just the destruction of forests on the grand scale which is having such a detrimental effect on the environment. In some parts of the world the planting of new trees in the name of progress is having a disastrous effect locally on existing wildlife and on local people. The 1989 BBC television series *The State of Europe* highlighted one example in Portugal where money from the European Community is subsidising large plantations of eucalyptus in the formerly unspoilt area of central Portugal known as the Alentejo, a unique feature of which are the 'Montadas' dominated by cork and holm oak trees. It is a harsh area, one of the hottest and driest in Europe. It is home for some rare wildlife – animals like the lynx, and birds like the Montague harrier and the great bustard. Now an Australian immigrant – the eucalyptus – is being planted to provide wood pulp for paper. The eucalyptus sucks the soil dry and robs it of the nutrients on which other plants depend. As the land dries the traditional farmers move out – and so does the wildlife. The justification in Brussels is that the country is poor and needs the income provided by the trees – but at what cost to the environment?

SAVING PLANTS IN BRITAIN

The International Union for the Conservation of Nature (a union of 62 governments, government agencies and non-government groups totalling 664 members dedicated to saving species) has a Conservation Monitoring Unit, based at Kew Gardens, which has recorded some 25 000 species world wide which could become extinct. Most of them are in the tropics where botanists, too, are a rare breed. Without the scientists to do the work of collecting and cataloguing them it is all too common for many rare and precious plants to simply disappear. It is estimated that fewer than 1 per cent of the world's plant species have been studied. If present trends continue it is thought that some 60 000 different plant species will be lost before the middle of the next century.

In order to help conserve the world's wild plants the Botanic Gardens Conservation Strategy was launched in December 1989. There are some 1500 botanic gardens and arboreta around the world and the idea is that they should get together and cooperate on saving threatened wild species – particularly those species used in medicine, as well as fruits, vegetables and spices. The Strategy aims to achieve some order in plant conservation and put it on a scientific and technical footing.

One initiative to save plants was begun in Britain in 1989 with the launch of an organisation called Plantlife. The figures it uses to support its campaign emphasise that it is not just in the tropics that plants are under threat. Plantlife has started its fight to save plants by focusing on what is happening here in Britain, because it is not just the burning of forests which is causing havoc with ecosystems and destroying habitats. Britain is perhaps one of the most enlightened countries in the world as far as conservation is concerned, yet the damage continues.

Britain has lost 97 per cent of its wildflower meadows, 190 000 miles of hedgerows have been lost – enough to circle the earth seven times – and three-quarters of Britain's heaths have been lost. Perhaps one of the saddest aspects of the problem in Britain is that more than half Britain's natural peat bogs have been lost and the rest are under threat, partly because people are digging it up to sell for use on gardens so that cultivated plants can grow better! In **Scotland**, ill-conceived forestry stimulated by tax avoidance schemes is destroying much of one of the largest areas of natural peat bogs on earth. By 1988 conifers covered 148 000 acres of peat bogs in east Sutherland and Caithness. Though there has been some recognition of the importance of this area, forestry is to be allowed to continue over at least another 100 000 acres. Plantlife believes what is happening in this so-called 'Flow Country' is a disaster.

Because of the threat to Britain's peat bogs scientists from the Nature Conservancy Council recently carried out a survey of 120 peatlands in **Wales** and were surprised by what they found. They discovered a previously unknown species of fly, six species of fly never before recorded in Britain and more than 100 other species recorded for the first time in Wales including beetles, butterflies, plant bugs and spiders. If these sort of discoveries can be made on our own doorstep it illustrates well that there must be thousands of other unique species yet to be discovered in more remote parts of the world.

Other well-known areas of natural beauty in Britain are under threat. They include the **New Forest**, where pressures from recreational use are causing severe damage. Acid rain is having an effect on the trees. Grazing by ponies and cattle, which keeps the forest open allowing many rare plants to survive, is declining because the 'Commoners' (the people who look after the ponies) are facing a crisis. The market price of ponies has not increased since 1975 and more ponies are being killed by traffic on the roads – 200 a year. In addition, the price of houses is so high now that only wealthy people are buying them and they are not taking up the Commoners' rights. In the world context it may not be in the same league as the destruction of the rain forests, but it is a typical example of the increasing pressures on ecosystems.

The **Cairngorms**, too, are under threat. There many arctic and alpine plants have survived since the Ice Age. Rare species include the Arctic Mouse-ear, Alpine Foxtail, Tufted and Highland Saxifrages and the Hare's-foot Sedge. Now the Cairngorms may be further damaged by those who want to extend the skiing slopes and build a new access road.

The **Norfolk Broads**, another popular area, are also threatened. Pollution from fertilisers and sewage has meant that the sight of water lilies floating on clear waters is now but a memory. The rare Fen Orchid grew in ten sites in 1959, now it is confined to three, with fewer than 100 plants found in the summer of 1989. And pressure from the building of new roads is adding to the destruction of ancient woodlands, putting more plants and habitats under threat.

Since records began, no fewer than 22 species of plants have been lost from Britain – plants like the Marsh Fleawort, which became extinct in the late nineteenth century when its wetland habitat was drained, and the Blue Iris, which was lost in 1982 following wanton destruction by a farmer.

Many countries now have Red Data books listing species in danger. Britain's Red Data book lists 317 threatened species of flowering plants, of which 47 occur in only one place and 31 in only two places. Plantlife cites the Cornish Hybrid Heather, of which only two or three bushes remain in the world, growing naturally on the Lizard. Plantlife hopes to campaign to protect habitats under threat, to restore devastated chalk grasslands and to lobby for more effective laws to protect plants and habitats initially in Britain but eventually, with other organisations, world wide.

One depressing aspect of the battle to conserve habitats is the apparent ease with which sites in Britain designated as Sites of Special Scientific Interest (or SSSIs) can be exploited and their special status overridden. In the year ending March 1989, of 4846 SSSIs covering 1.4 million hectares 241 had been lost or damaged in some way – 80 more than in the previous year. Some of the damage was short-term, but there were 39 instances of long-term damage being caused to 1500 hectares. Only about 10 per cent of the damage was caused by agriculture. Most of it was due to activities like peat digging and road and house building.

Beyond our shores there is a similar story of destruction in many countries. In northern Greece, for example, in an effort to bring prosperity to the Gulf of Amvrakikos, the European Community is paying for six fish farms in the area. Unfortunately those in Brussels who made this decision seem not to have taken any notice of the environmentalists along the corridor in the same organisation who point to environmental directives which forbid the building of fish farms in this area.

The Gulf of Amvrakikos is a wetland area of international importance. It is a complex system of lagoons, marshes and rivers constituting one of the rarest habitats in the world. The fish farms draw millions of gallons of water from the wetlands and pump back huge quantities of fish waste. Traditional fishermen are finding their catch dwindling.

While the preservation of plantlife is essential, of greater public appeal are campaigns aimed at preserving animal species. The destruction of the environment world wide and the exploitation of animals by man has led to many animal species being threatened with extinction.

The campaigns to save threatened species have focused on a few spectacular animals and their stories are a sad reflection on man's impact on the environment.

* * *

THE ELEPHANT

Perhaps the most publicised of threatened animals recently has been the elephant. Due to the illegal trade in ivory there has been a dramatic decline in the number of elephants. Across Africa as a whole in 1979 there were 1.34 million elephants. Today there are just over 600 000. That blanket figure conceals differences between individual African countries. The decline has been particularly bad in East Africa where poachers armed with AK 47 rifles and chain saws have drastically reduced the population.

What has been particularly worrying is that females are not breeding as fast as they used to. This is thought to be because females prefer mature males of about 30 years old. Yet it is just these males which are sought after by poachers, as they have the biggest tusks.

Even elephants in wildlife reserves are not safe. Kenya, for example, has lost some 70 per cent of the elephants living within park boundaries.

Because of the perilous situation, there were many calls for a worldwide ban on the trade in ivory. This was not always greeted with universal approval, some people claiming a worldwide ban would simply drive the trade underground and drive up prices. There was much criticism of the organisation known as CITES (the Convention on International Trade in Endangered Species). Since 1977 the African elephant has been listed in Appendix II of the Convention as an animal in which trade is allowed under some circumstances. But in 1989 seven countries, led by Tanzania, proposed that elephants should be transferred to Appendix I – the list of animals in which trade is banned. Other countries in southern Africa argued for a form of agreement which would permit limited trade in ivory. This was because in southern Africa, where there is little poaching, and elephant populations are carefully managed, the number of elephants has actually been going up. In South Africa, for example, the tiny population of 7800 elephants has increased to 8200 in the last ten years. In Zimbabwe the figure was 30 000 ten years ago compared with 43 000 today, and in Botswana the elephant population has gone up from 20 000 in 1979 to 51 000 today. The debate over what should be done to save the elephants reached an acrimonious climax in October 1989 amid accusations that CITES had been hijacked by the ivory traders. CITES argued that limited trade in ivory should be allowed to continue. But commonsense prevailed and at this crucial meeting countries voted 76 to 11 with four abstentions in favour of a worldwide ban on the trade in ivory. The delegates also voted in favour of a scientific panel to be set up to decide when and if certain African countries can resume trading, in the event of their being able to demonstrate that they have a self-sustaining stock of elephants. The agreement was hailed by the British representative at the meeting as 'a very important decision – one that means real hope for the future of the African elephant'. Unfortunately, the British Government undermined faith in its commitment to the ivory ban when it later allowed Hong Kong ivory traders to sell their stockpile of nearly 700 tonnes over a six-month period. Within a very short time it was said the poachers were back in Kenya. According to Richard Leakey, the Director of the Kenya Wildlife Service, while in the last six months of 1989 the number of elephants killed had decreased to between 15 and 20 a month, that figure rose to 50 in December 1989 and in the first few months of 1990 reached 100. Captured ivory poachers said that the traders had returned to East Africa because they believed the market had opened up again in Hong Kong!

The six-month 'grace' period for Hong Kong ended in July 1990 but there was still a large stockpile of ivory remaining unsold and Hong Kong brought in controls to prevent it leaking on to the market. The collapse of the ivory market has had severe repercussions for the 3000 people in Hong Kong who made their living carving or selling it. According to their spokesman, tourists have turned against ivory and regard them now as murderers.

The ban on the trade in ivory, though welcomed by most countries, may still not be sufficient to save the elephant if experience is anything to go by. In 1976 CITES banned all trade in another endangered African species, the black rhino. There were 50 000 of them then – today there are just 3500. They are shot by poachers for their horns which are made into dagger handles in the Yemen and ground into exotic medicines in the Far East. Rhino horn commands a higher price than even gold.

THE WHALE

There can be few people now who have not heard of the campaign to 'stop the bloody whaling'. It has been one of the most effective of all environmental campaigns, spearheaded by the organisation Greenpeace. Commercial whaling was at its peak in the 1960s when whales were being killed at the staggering rate of 70 000 a year. The International Whaling Commission was formed in 1946 to protect whale species facing extinction. In 1986 it agreed to a moratorium on commercial whaling until 1990 when the situation would be reviewed. But there were certain exclusion clauses which allowed whales to be caught for scientific purposes or for 'subsistence, cultural and social reasons' – a clause which has been exploited by Iceland, Norway and particularly Japan. In 1989 Japan issued permits for the killing of 300 minke whales, Iceland for 78 and Norway for 30.

At the 1989 International Whaling Commission annual meeting in San Diego conservationists lobbied to get an extension of the moratorium on commercial whaling after 1990. The Commission heard evidence that several whale species were far nearer extinction than had been suggested up till then. Originally there were 250 000 blue whales in the ocean. In 1960, when the blue whale became protected, there were only 12 000 left. The scientists suggested the figure was now down to 500. Fin whales had declined from 500 000 originally down to 100 000 in 1976 and by 1989 had reached between 2000 and 5000, while sperm whales had come down from 1 million originally to 20 000. The Japanese disputed these figures. They maintain whaling is important to estimate the age and populations of whales and they say the sale of the whale meat also raises funds. Conservationists dismiss the arguments and claim 'scientific' whaling is simply commercial whaling under another guise.

THE CHIMPANZEE

Chimpanzees are another species which, partly because of their great appeal, are threatened. With major changes in the landscape of Africa in recent years, particularly in West and East Africa, at least two sub-species are on the endangered list. Chimpanzees have disappeared from Benin, Gambia, Togo, and Burkina Faso. And they are expected to disappear soon from Guinea Bissau, Ghana, Nigeria, Burundi and Rwanda. Not more than five African countries now support populations in excess of 5000. With these animals it is not just the destruction of their habitats which is causing their decline: female chimpanzees are often shot so that their babies can be collected and sold for the entertainment trade. One of the world's leading experts on chimpanzees, Jane Goodall, describes their situation as desperate. She has set up an institute to improve research and to help protect them.

THE PARROT

Habitat destruction and the trade in wild birds has led to many species of parrots facing extinction. The plight of the parrot constitutes a wildlife crisis. One hundred species – nearly a third of the world's total – are at risk, and 77 are in immediate danger of extinction. The parrot is a bird universally loved for its bright plumage, its powers of mimicry and the fact that it can so easily be tamed. Yet their very appeal to man is part of their undoing. Parrots have been on earth ten times longer than man and it is a tragedy that their continued existence is threatened.

In particular danger due to the illegal trade are the macaws and amazons of the Americas and the cockatoos of Indonesia and the Philippines. In the Caribbean and continental South America there are 40 species in real danger. And there are other threatened species in New Zealand, Australia, South East Asia, India and Africa.

One difficulty in trying to protect parrots stems from the nature of CITES. The Convention requires that all import and export of live animals which are under threat should have a licence. But to qualify, the species has to be seen to be threatened. The problem is that very little is known about some parrot species and without adequate knowledge of them in the wild it may not be possible to prove a species is endangered until it is too late. The other problem is that even those parrots which are protected are caught and sold illegally. The Lear's macaw and the Spix's macaw from Brazil are both protected by CITES and by Brazilian law, yet they have been hunted to the brink of extinction.

In 1989 the International Council for Bird Preservation launched a campaign to protect the parrot. Among other things it drew attention to the appalling way parrots are trapped and then transported to the world's pet shops. Thousands die in transit. For every parrot which gets to a pet shop another four have died on the way. The campaign called for a ban on the import of all threatened parrots into the European Community, new regulations governing the conditions in which parrots are kept during transport, a campaign to persuade people who buy parrots to make sure they have been bred in captivity and a tax on the sale of parrots to raise funds to preserve the parrots' habitats.

ANTARCTICA

The threat to thousands of plants and animals is mainly due to the encroachment by man on their habitats. But there is one remaining place in the world where man's impact has so far been marginal – the Antarctic. Here the bitter cold has kept mankind at bay – and many want that situation to continue.

Antarctica has often been called the world's last great wilderness. It contains 10 per cent of the world's landmass and 90 per cent of the world's ice. The Southern Ocean contains an abundance of wildlife including whales, seals and birds.

Most people want to see Antarctica preserved in one way or another. But there is much controversy as to how that can best be done. There are those who claim that any exploitation of the Antarctic's mineral reserves will be a disaster for the delicate ecosystems of the area. Others claim that an orderly and controlled exploitation is the best way forward and would ensure that the vast majority of the Antarctic land mass would thus be conserved.

The history of man's involvement in the Antarctic is relatively short. Conflicting territorial claims over the Antarctic threatened to lead to conflict until the scientific programme known as the International Geophysical Year (1957–8) led to a defusing of tensions and international cooperation by scientists of 12 nations, all with an interest in the area. Out of this international cooperation grew the Antarctic Treaty, signed in 1959 by 12 nations: Argentina, Australia, Belgium, Chile, France, Japan, New Zealand, Norway, South Africa, the United Kingdom, the USA and the USSR. This Treaty froze existing territorial claims and prohibited future ones. Its aim was to ensure that the Antarctic was only to be used for peaceful purposes. It set the scene for increased scientific research and cooperation. Issues relating to the Antarctic are discussed at two-yearly meetings of the twelve States. Any decisions have to be ratified by the countries and no country can be bound by a decision without its consent. That means fast decisions on controversial issues are unlikely. Because the Antarctic Treaty does not cover the exploitation of resources, other conventions have been agreed. These include a convention on the conservation of Antarctic Fauna and Flora, the Convention for the Conservation of Antarctic Seals, agreed in 1972, and a convention on the conservation of Antarctic Marine Living Resources agreed in 1980.

The most controversial issue of late has been the prospects of mineral exploitation in the Antarctic. In 1977 the 12 countries agreed on a temporary moratorium on mineral exploitation. Then in 1988 an agreement for controlled exploitation of mineral resources was reached but in 1989 both the Australian and French Governments decided they did not want any exploitation of minerals in the area and refused to sign. Instead they want the Antarctic to be declared an International Wilderness Reserve where minerals activities are banned. Their view is supported by Belgium and Italy. Without the support of Australia and France, both of whom claim territory, the agreement cannot come into force and mineral exploitation is effectively prevented.

With the world suddenly more environmentally conscious, various options for the further control of activities in Antarctica were due to be discussed in 1990 at a Special Antarctic Treaty Consultative Meeting. One option suggested by New Zealand fifteen years ago is to make Antarctica a World Park. And Australia has proposed that the Treaty countries negotiate an 'Environmental Convention' similar to the World Park concept which would regulate all human activities and prevent minerals mining.

Despite the promising start which has been made over the decades towards conserving Antarctica, there are already signs that commercial pressures are beginning to damage the ecosystem, with overfishing one of the most threatening. Large-scale trawling began in the 1960s. In 1971 Eastern European bloc countries were catching 400 000 tonnes of fish a year. This is believed to have led to a decline in the number of fish, with the result that the fish catch has decreased substantially. Fishing for the Antarctic Krill, a shrimp-like animal, peaked in 1981–2 when 528 000 tonnes were caught mainly by Japan and the Soviet Union.

In 1980 the Convention on the Conservation of Marine Living Resources was signed to control fishing. But it has not succeeded in stopping overfishing because countries catching the fish are not required to provide figures showing how much fish they are taking out.

CHAPTER THIRTEEN

Prognosis

Having reviewed some of the significant threats to the environment, the question has to be posed: is it too late to save the planet? It sounds a dramatic question – some might well say over-dramatic – but the evidence is overwhelming that the world cannot simply continue on the present path without ending in chaos. In the last few years many governments have woken up to the threat, and many international meetings have concluded with ringing declarations. A start has been made at reversing some of the undesirable trends – though only a very tentative one.

Even given firm commitments to a change in direction, it will take many years before some of the more serious threats are reversed, such is the momentum that has now been established. The greenhouse effect and ozone depletion will take many decades to stabilise, given that much of the damage to be expected in the near future has already been set in train.

Some threats can be dealt with more easily, however. The ban on the trade in ivory, for example, is likely to have a fairly immediate effect in helping to protect the elephant. But the shrinking forests and the eroding soils cannot be replaced overnight, whatever decisions are taken now. And, meanwhile, the world's population goes on growing – by 84 million a year.

On the question of the world's food supply there is some evidence that the large increases which have been made possible by modern technology may well have peaked. Between 1950 and 1984 the world's output of grain increased 2.6 times. But since 1984 there has been little increase. People in Africa and Latin America now have less to eat than they did ten years ago and infant mortality is increasing. Forty thousand babies and young children die each day in the developing world. In Africa there are now fewer cattle per head of population than in 1950, and grasslands are fast losing their ability to sustain the cattle herd, due, among other things, to soil erosion.

The world's environmental assets have not figured properly in economic accounting. The measure of a country's Gross Domestic Product takes account of the total of the goods and services a country produces, minus a figure for the depreciation of its plant and equipment, but not its environmental assets – neither non-renewable resources like oil and coal nor the renewable assets like forests. Yet if some account is not taken of environmental assets then the GDP presents a grossly distorted indication of a country's long-term prospects, and gives no indication of its potential for sustainable development.

Putting a monetary value on the preservation of the climate is particularly difficult. A carbon tax is one way to encourage a policy of energy conservation and the development of renewable energy sources which will help reduce the rate at which the climate will change.

Another scheme to help control carbon emissions was suggested by the Americans at a meeting of the Intergovernmental Panel on Climate Change in December 1989. It involves establishing a system of emission credits for all greenhouse gases. Individual countries would be allowed to emit up to a stated amount of carbon each year and these credits could be traded. That means that a country which wants to emit more than its quota could buy emission credits from a country which could manage by emitting less. Such a system of emission credits has been branded by some as unworkable internationally. How would the credits be set in the first place? How would a country's compliance be monitored and how would the system be policed? Some have seen the American suggestion as at least a sign that the USA acknowledges the need for something to be done. Others believe that politics is driving the proposal.

The likelihood that global warming is a real risk is becoming increasingly certain the more scientists look into the issue. The British-led panel which examined the scientific basis for the concern as part of the preparations for the Second World Climate Conference came to the conclusion that if the world continues to emit carbon at the current rate then by the year 2030 the average global temperature will have increased by between 2 and 4°C. But, while many countries recognise the seriousness of the threat, there are still people who feel it is being exaggerated. Much of industry in the United States has adopted a defensive attitude, whereas in Britain the CBI has on the whole been fairly positive. If some sort of carbon tax to limit emissions were to be agreed, some claim it would have to be introduced internationally. Others feel that that would not be necessary as energy costs already differ considerably between different countries.

The development of an international convention to protect the climate, though most countries recognise the need for one, will be very difficult. Experts are predicting considerable problems in getting everyone to sign. Developing countries are particularly worried, firstly because there are so few experts who can advise them, and secondly because there is a fear that such a convention may well have the effect of slowing down their development. On so many environmental issues it is the industrialised West which has caused the problems and, in the eyes of some, it is up to those in the West to put things right.

This is an attitude common to a number of environmental issues, not just global warming. With the ozone layer, for example, the developed world has caused most of the problem, and the extra UV light which may fall on earth as a result of the depletion of the ozone layer has led to an attitude among some people in the developing world of 'hole in sky, white man fry'. In the equatorial regions where mankind has developed his own pigmentation to counteract the natural effect of the sun, why should they worry too much?

But in order to work, international conventions on the environment – especially those addressing the subject of global warming – have to have the support of the developing world and particularly of countries like India, China and Brazil which contain such a large proportion of the world's population.

Man is essentially short-sighted but he is totally blind if he thinks that nature can be controlled. There is increasing evidence that the balance of nature is now being upset in a major way and man must now recognise this. The planet earth has seen many catastrophes in its long history. Until now they have mostly been of natural origin. But in the comparatively short time mankind has trod the earth he has set in train trends which will be difficult, perhaps impossible, to reverse.

Information for this book came from a variety of sources including the following:

The Worldwatch Institute's reports:

'Slowing Global Warming – A Worldwide Strategy' – October 1989

'Renewable Energy: Today's Contribution, Tomorrow's Promise' – January 1988

'Conserving Water – the Untapped Alternative' – September 1985

'Water for Agriculture – Facing the Limits' – December 1989

'Rethinking the Role of the Automobile' – June 1988

'Clearing the Air – a Global Agenda' – January 1990

'The Greenhouse Effect – a Scientific Basis for Policy' – The Royal Society, 1989

Reports from the House of Commons Energy Committee

Reports of the Committee on the Medical Aspects of Radiation in the Environment, 1989

Report from the UK Stratospheric Ozone Review Group (HMSO), 1988

Tenth Report of the Royal Commission on the Environment 'Tackling Pollution, Experience and Prospects' (HMSO), 1984

'Acid Rain – The Ecological Imperative for Pollution Controls', published by the World Wide Fund for Nature, 1986

Digest of Environmental Protection and Water Statistics No. 11 (HMSO), 1988

Rainforests, Land Use Options for Amazonia with introduction written by Dr Norman Myers. Published by Oxford University Press and the World Wide Fund for Nature, 1989

The Ages of Gaia by James Lovelock, published by Oxford University Press, 1988

Blueprint for a Green Economy, by David Pearce, Anil Markandya and Edward B. Barbier, a report prepared for the UK Department of the Environment, published by Earthscan Publications Ltd

Various reports on many subjects for Friends of the Earth and Greenpeace; the World Wide Fund for Nature; Plantlife; the Royal Institute of International Affairs (Greenhouse Effect): The Panos Institute; The World Health Organisation; the UK Hazardous Wastes Inspectorate; The House of Commons Environment Committee; The Ministry of Agriculture, Fisheries and Food; The Department of the Environment; the Department of Energy; the Atomic Energy Authority; Nirex; United Nations Environment Programme; The Association for the Conservation of Energy; *New Scientist*; *Scientific American*; *Science.*

Index

INDEX

INDEX